MINERS
AND THE GREAT WAR

Private (Piper) Henry J. Irving: 20th (1st Tyneside Scottish) Battalion, Northumberland Fusiliers. Unlike many others of the 20th he survived the carnage of 1 July 1916, only to be wounded at the end of the following month, on front-line duty with his much-depleted battalion. From a Northumberland mining family, Henry had migrated from Ashington to Ryton, County Durham, finding employment at Greenside Colliery. Aged twenty-four, he swapped his pick for a rifle on 28 October 1914, remaining in service until discharge on 7 March 1918. Michael Hardy

MINERS
AND THE GREAT WAR

Brian Elliott

AN IMPRINT OF PEN & SWORD BOOKS LTD.
YORKSHIRE - PHILADELPHIA

For my grandchildren
Edie, Ellis, Harri and Rhys

First published in Great Britain in 2025
by Pen & Sword Books Yorkshire - Philadelphia

Copyright © Brian Elliott, 2025

ISBN 9781473827264

The right of Brian Elliott to be identified as author of this work has been asserted by him in accordance with the Copyright, Designs and Patents Act 1988. A CIP catalogue record for this book is available from the British Library. All rights reserved. No part of this book may be reproduced or transmitted in any form or by any means, electronic or mechanical including photocopying, recording or by any information storage and retrieval system, without permission from the Publisher in writing.

Printed by CPI UK

Design: Paul Wilkinson

The Publisher's authorised representative in the EU for product safety is Authorised Rep Compliance Ltd., Ground Floor, 71 Lower Baggot Street, Dublin D02 P593, Ireland. **www.arccompliance.com**

Pen & Sword Books Ltd. incorporates the imprints of Pen & Sword Books: After the Battle, Archaeology, Atlas, Aviation, Battleground, Discovery, Family History, History, Maritime, Military, Politics, Select, Transport, True Crime, Fiction, Frontline Books, Leo Cooper, Praetorian Press, Seaforth Publishing, Wharncliffe and White Owl.

For a complete list of Pen & Sword titles please contact
PEN & SWORD BOOKS LIMITED
George House, Beevor Street, Off Pontefract Road, Hoyle Mill, Barnsley, South Yorkshire, England, S71 1HN.
E-mail: enquiries@pen-and-sword.co.uk
Website: www.pen-and-sword.co.uk
or
PEN AND SWORD BOOKS
1950 Lawrence Rd, Havertown, PA 19083, USA
E-mail: uspen-and-sword@casematepublishers.com
website: www.penandswordbooks.com

CONTENTS

Foreword by Sir Michael Parkinson 6
Preface 8
Abbreviations 10

Chapter One	**King Coal**	11
Chapter Two	**1914: Pit Duds to Khaki**	18
Chapter Three	**1915: Innocence Lost**	41
Chapter Four	**Silverwood: A Wartime Colliery**	82
Chapter Five	**1916: Unspeakable Horror**	106
Chapter Six	**'Conchies': The Men Who Said No**	144
Chapter Seven	**1917: Stinking Trenches and Surreal Landscapes**	153
Chapter Eight	**Behind and Beyond the Wire**	190
Chapter Nine	**1918: Dark Pits of War No More**	204
Chapter Ten	**Demobbed and Demoralised**	231
Chapter Eleven	**In Memoriam**	246

Timeline 259
Acknowledgements 262
References 263
Bibliography 274
Index 278

FOREWORD

By Sir Michael Parkinson

I am the son of a miner. I never fully experienced what a dirty and dangerous job my father did for a living because he was determined that I didn't follow him underground. He once took me into Grimethorpe pit and showed me where men worked half naked on their bellies in hellish heat, picking away at a coal seam with hand tools. It was a like a scene from a horror movie. My father said nothing, he didn't have to, I was scared to death. On the way back he broke the silence but only to gently warn me that if he  ever saw me at the pit gates he'd 'kick my arse all the way home'. He was twelve when he first went to work. When he took me down the pit I was fourteen, a Grammar School boy with a headful of Hollywood movies and Len Hutton cover drives. Thank God he dissuaded me, I wouldn't have lasted one shift.

My father died too early but not before he saw that his lesson as my careers adviser had borne fruit. He was a victim like many of his generation of the medieval working conditions that were allowed to exist for too long by an unspoken conspiracy between uncaring, greedy pit owners and politicians who were too worried about the nation's balance sheet and too attached to a class system, that allotted a man's position in society according to the circumstances of their birth, to try and improve what was happening beneath their well-shod feet. As a result, when in 1914 Kitchener asked every man to do his duty, for many miners it was an easy choice to replace another poorly paid shift in an underground death-trap for the machine guns and annihilating artillery of the German Army on the Western Front. They left the mines in their droves. By the summer of 1915, over a quarter of a million miners had joined the military and the mark they left on the war effort was indelible and unique.

Brian's book is a fascinating study of this little-known aspect of that terrible conflict. I first came across the miners' contribution in Sebastian Faulks's wonderful book, *Birdsong*. Its depiction of the subterranean war conducted on the Western Front by English and German miners and their attempts to tunnel beneath each other's lines was horrific, compelling and fascinating. Whilst this was the most obvious aspect of the miners' contribution, Brian also reveals that they served in all the theatres of the war and in all branches of the military.

Miners became prized recruits because of their fitness, strength, durability and training in First Aid and also because of their unity as a group, their shared idea of never leaving a workmate behind forged in the unforgiving workplace where they plied their trade. There is also the fact that they came from an environment that produced a particular type of person. I was a product of a pit village and it creates within you a sense of community, a bloody-mindedness and a suspicion of those in authority and indeed anyone from the outside which would've made miners fine comrades and formidable foes. It is no surprise therefore, that miners received a huge number of gallantry awards including Victoria Crosses.

This is an important history of a mostly overlooked, disregarded group who have regularly been the subject of hostile criticism and outright dislike. By taking these men out of the darkness and showing their faces to the world, Brian has demonstrated what an extraordinary bunch of people they were and remain so to this day. I should know, I grew up amongst them.

Hewers stripped to the waist undercutting a seam of coal. Brian Elliott

PREFACE

From underage teenagers to over sixty veterans, about half a million miners (aggregated number) served in the Great War, mostly as volunteers, though large numbers were also transferred or directly targeted from collieries for their specialist skills, most notably for the new tunnelling companies of the Royal Engineers (RE). Miners with first aid and mine rescue experience were also highly prized recruits for the RE, and for service with the Royal Army Medical Corps (RAMC) and various 'medical' units of other regiments. Former pit pony drivers, 'horsemen' and colliery blacksmiths, also had obvious advantages for the Army's logistical requirements. Smaller miners were welcomed into the Bantam regiments, though this may have facilitated even more underage enlistments. Miners' mutual comradeship at times of great danger was an obvious advantage for the military, Haig and most commanders were well aware of this generational asset.

Although there was a high infantry concentration, miners were present in all of the military services and theatres of war, including the new, fast developing air force.

The maturity and the bravery of miners on and under the land, at sea and in the air was truly extraordinary, many of them still in or barely out of their teens. Thousands of honours were awarded including many DCMs (Distinguished Conduct Medals) and more than fifty VCs (Victoria Crosses).

Regions varied, but as many as one in three of a colliery's employees volunteered for military service during 1914-15, especially for the new Pals battalions but also for a variety of regiments in coalfield regions, from Scotland and North East England through Yorkshire and the Midlands to South Wales, Somerset and Kent. Some, such as the Durham Light Infantry, had a huge proportion (c.85%) of miners in their ranks.

At the same time, the Army and Royal Navy wanted more and more coal for the war effort, placing dual demands on the 'stay at home miners' and on the managers and owners, who tried to maintain and increase production with cohorts of old, very young and the inexperienced. More girls and women were recruited to work on pit tops in some areas, despite opposition from the miners' unions.

During the Great War almost 7,000 miners were *killed at work*, mostly in everyday accidents such as roof-falls. Countless, mostly unrecorded, others suffered due to long-term injuries and lifelong ailments.

The book is wide ranging, sampling all the coalfields of Britain during and after the war, and in part is based on hundreds of 'mini-biogs', many of them never published before. It also pays homage to the relatively small but significant voices of miners who had the courage to claim exemption or 'special treatment' on the grounds of conscience: the conscientious objectors (COs). Another somewhat neglect area is also featured: miners as prisoners of war (PoWs).

Postwar, despite political assurances and promises, many of the miners returning or attempting to return to the their pits faced growing poverty, even destitution, especially those with mental and physical disabilities. As will be shown, the 'return to the pits' assurances did not work for all due to many collieries not having the capacity for re-employment; and the fact that so many 'miner-soldiers' discharged as 'fit and well' in reality suffered a variety of physical and mental ailments, often emerging in later life. Mining families' demises were not helped by the impact of the influenza pandemic which raced through close-knit pit villages and coal towns.

Disputes, strikes and lockouts permeated most coalfields during the early 1920s at a time when colliery war memorials and tribute medals were created in honour of the living and the dead. 'In the trenches one minute and on the dole the next' was a Labour Party election poster not far from the truth. Those that did work had had their wages slashed by about a third or more, preluding a series of hunger marches of the 1930s. Industrial towns such as Barnsley were especially hard hit, which is why Eric Blair (George Orwell) came to the town as part of his research for his book *The Road to Wigan Pier*, experiencing first-hand what life was like for a miner and his family.

Thus, a large and varied range of sources have been used in the book, including hundreds of case studies of 'miner-soldiers and sailors', many of them the otherwise 'lost' or 'forgotten voices' of the war.

My acknowledgements are itemised at the end of the book, but here I would like to pay special tribute to the late Sir Michael Parkinson for providing me with a heartfelt and most appropriate Foreword. And also to my publishers Pen and Sword for their immense patience and kindness to me over eleven years of research and writing.

Abbreviations

ASC	Army Service Corps
BEF	British Expeditionary Force
DCM	Distinguished Conduct Medal
DLI	Durham Light Infantry
DMA	Durham Miners Association
DMC	Dalton Main Company/Collieries Ltd
DSM	Distinguished Service Medal
DSO	Distinguished Service Order
CLC	Central Labour College
CMOC	Coal Mining Organisation Committee
CO	Conscientious Objector
CWGC	Commonwealth War Graves Commission
DoRA	Defence of the Realm Act
FA	Field Ambulance
FAU	Friends Ambulance Unit
HOS	Home Office Scheme (aka Brace Scheme)
ILP	Independent Labour Party
KOYLI	Kings Own Yorkshire Light Infantry
MFGB	Miners Federation of Great Britain
MGC	Machine Gun Corps
MM	Military Medal
MST	Military Service Tribunal
NCAC	National Council Against Conscription
NCC	Non-Combat Corps
NCF	No-Conscription Fellowship
NCO	Non-Commissioned Officer
NEIMME	North of England Institute of Mining and Mechanical Engineers
NMA	Northumberland Miners' Association
NUM	National Union of Mineworkers
PoW	Prisoner of War
PRBCO	Pearce Register of British Conscientious Objectors
RAMC	Royal Army Medical Corps
RN	Royal Navy
RNVR	Royal Navy Voluntary Reserve
SCBA	Self-Contained Breathing Apparatus
The Fed	South Wales Miners Federation
TUC	Trades Union Congress
VC	Victoria Cross
WEA	Workers Education Association
WNI	Work of National Importance
Y&L	York and Lancaster Regiment
YMA	Yorkshire Miners Association

Chapter One

KING COAL

The coalminers of Great Britain made an enormous contribution to the war effort during the years 1914 to 1918. Great waves of young pitmen rushed to enlist, 191,170 of them responding to Kitchener's call to arms within seven months of the start of hostilities, or about one in five of all mining personnel (men, boys and women). By the summer of 1915, 250,750 workers had left their collieries for military service.[1] Some mines were so seriously depleted of labour that a two-shift system had to be introduced in order to keep production going, a few pits closed for a while, and within weeks of war, many of the larger collieries had lost more than half of their key workers. Those that remained – a great 'Underground Army' – worked extra hard to produce the coal that the country demanded to keep the home fires burning and drive a huge military machine.

Coal was ubiquitous in 1914. It was everywhere. It fuelled steel making, shipbuilding and ships, and just about all of industrial Britain. It generated new electricity, transported people and heated homes. No coal equalled no power, no life. Thus the First World War became the world's first great industrial war.

The coal-getters themselves were often a downtrodden and much-maligned group of workers. But through their umbrella union the Miners' Federation of Great Britain (MFGB) they always stood up against injustices. The nation and politicians thus shuddered in 1912 when over a million miners downed tools nationwide for a strike that was euphemistically known as a 'holiday' so as to avoid legal intervention by the Government. At last a minimum wage was agreed instead of the divisive 'Price Lists', pay rates that might vary from one colliery to another and one region to another. However, wage complexities and exploitation of their labour force by some of the owners and numerous health and safety issues remained.

In the coalfields there existed a huge 'invisible' but vital workforce. The girls and women in mining families, often running large households that might include a lodger or two, were absolutely indispensable, the engines that kept everything moving. Some females even worked on the surface of mines, as hauliers, pushing wheeled tubs to the mechanical tipplers that upturned the coal into bunkers or on to belts, or picking dirt and stone from coal on the screens. It was hard, it was dirty and it was dangerous work. Most regional miners' unions objected to female employment but in parts of Lancashire, Cumbria and south Wales there was a history of 'pit brow lasses'. Women

'Pit Brow Girls' portrayed on a postcard sent from Wigan in May 1914. Brian Elliott

and girl 'miners' were viewed as 'curiosities' by some local photographers who realised the commercial advantages of their studio images portrayed on carte de visites (small card-backed photographs), cabinet photographs (larger card-backed images) and popular postcards.

Addressing delegates at the first National Mining Conference on 29 July 1915, the Minister of Munitions, David Lloyd George, spoke with great passion about the indispensability of 'King Coal' in times of peace and war. Without the black gold from the hills and valleys of the British coalfields there would be no victory against Germany.[2] This from a man who turned down the miners' demand for the coal industry to remain nationalised after the war.

The number of persons employed in collieries had reached seven figures for the first time in 1909, peaking at 1.1 million four years later.[3] It was a huge workforce, geographically spread from Clackmannan and Fife in Scotland to south Wales and Bristol and Somerset in the south west. New, deep mines were also being developed in the county of Kent.

Ranked according to output,[4] the seven major regions of coal production at the eve of the war were:

1. South Wales
2. North East
3. Yorkshire
4. Scotland
5. East Midlands
6. Lancashire & Cheshire
7. West Midlands

All coalfields contributed in varying degrees to voluntary enlistment into the armed forces from August 1914, through to 1916. In the North-East Newcastle upon Tyne had the busiest recruiting office anywhere in Britain. Even small coalfields in Cumberland, north Wales and Somerset, responded to Kitchener's call to arms in large numbers.

The hewers who got the coal from the seams were an elite part of the underground workforce, usually working in pairs stripped to the waist, kneeling or laying on their sides in the thinner and 'middling' workings. Almost 70% of all coal production in Britain in 1913-14 was obtained via manual means. Tools were hand-held, roofs supported by props or short 'sprags' (lengths) of timber. It was a highly skilled as well as physically hard job, undertaken with only a short break during a seven or eight-hour shift, endured in dismal light. The use of coal-cutting machines, driven by compressed air or electricity was unevenly distributed. Scotland was one of the few regions where machines exceeded traditional methods, other areas adopting new technology according to conditions, profitability, and the attitude of the owners.

Pushing or 'tramming' tubs of coal along low roadways was one of the hardest of all underground jobs. Brian Elliott

Physicality also played a part in the underground transport of coal by an array of haulage workers, usually young miners known as trammers, drawers or putters, who pushed wagons along narrow rails from the coal workings to either pit bottoms or the entrances of drift mines. The use of horses and ponies meant that greater loads could be carried. In 1912 there were 70,000 horses in British mines or about one to every twenty miners.[5] Some of the pony-drivers were an asset for the army because of their experiences with horses, as were the ostlers and the blacksmiths. Mainly in larger collieries, roped or chained trains of trams or wagons were hauled by mechanical means, powered via compressed air or electricity. But the haulage lads still had to execute the linkages, a job where damaged or broken fingers and toes were common; and far worse, 'runaway' trams could resulted in horrendous injuries and fatalities.

About four out of every five miners – like my paternal grandfather, a hewer – continued working during the war, their labours contributing to the burgeoning demand for coal. As already stated, the most underrated people in the industry at the time, a great unpaid, unofficial workforce, were the girls and women, wives and daughters of miners who often brought up large families in impoverished circumstances. Although not agreed to by the miners' unions, more girls and women were urgently recruited for pit-top work because of the shortfall of labour, especially after 1916 when surface workers were not excluded from conscription.

Although patriotism played a part in recruitment, for many miners joining the Colours had many other advantages. Many pit lads simply downed their picks and shovels in unison with pals in order to 'see the world' via a short continental holiday, sadly unaware of what horrors lay ahead. Among integral incentives was the prospect of regular pay, decent work clothes and boots; and of course escape from what was a hellish and highly dangerous job. In 1913 alone, 1,753 miners were killed at work and 178,000 injured, staggering statistics. Over a thousand deaths a year meant that most mining families experienced some form of loss. Most of the fatalities occurred in everyday accidents involving a single or small number of men and boys, especially in roof falls and haulage mishaps, but they were not always reported in newspapers.[6]

It was the multiple-fatality colliery disasters that attracted widespread media coverage during the war. There were ten examples, each involving at least five deaths, the worst at Podmore Hall Colliery ('Minnie Pit') in Staffordshire, where 156 men and boys were killed.[7] Podmore inspired Wilfred Owen's composition *The Miners*, one of the few of his poems published before his own death on the Western Front in 1918.[8]

For many young miner recruits the slow and relentless prospect of industrial disease was delayed, even halted. Respiratory problems were common in older miners, though health and safety regulation in the 1911 Coal Mines Act did improve the working conditions for newer workers; but breathing in coal dust and deadly silica from stone, continued to affect underground and some pit-top workers very badly. Pneumoconiosis or 'black lung' was debilitating for many miners and dreadfully distressing for their families.[9]

Working in cramped conditions and making repetitive and arduous movements when getting the coal also resulted in a variety of physical ailments, as did having to put up with damp and wet conditions. Ironically, it was this activity that labelled miners as perfect for the new tunnelling companies of the Royal Engineers. The complete absence of sanitary facilities and having to walk home 'wearing pit muck' was a part of the everyday life of the miner – conditions not too distant from trench life.

As a fit and skilled workforce, used to dealing with adversity, and co-operative when under extreme pressure, miners became a special, preferred group in a variety of military contexts. In pioneer corps they had vital roles in creating and maintaining hundreds of miles of trenches. It has only been relatively recently, thanks to Sebastian Faulks's novel *Birdsong* and the work of military historians such as Peter Barton, that the role of miners as tunnellers in the secret but often spectacular 'underground war' against their German opponents has become more widely known.[10] Specially recruited for (or attached to) the Royal Engineers, these remarkable men showed exceptional bravery throughout the conflict, though only one miner-tunneller, William Hackett, was awarded the VC, sadly posthumously. Hackett's extreme and unselfish gallantry, at Givenchy, overrode the cloak of secrecy that clouded recognition of the extraordinary feats of many other miner-tunnellers of the Great War.[11]

Two aspects of mining life had a significant impact on the roles of miners in the War. Mine owners and groups of mine owners, and of course the miners themselves had a vested interest in First Aid. New charitable organisations such as the St John's Ambulance Association targeted mining areas in order to practice and demonstrate their skills and services. Instructors were also able to provide graded qualifications for the miners. A great deal of expertise was also developed thanks to a large number of local and regional First Aid competitions, medals and trophies in the form of cups and shields presented on a regular basis. Many of the 'ambulance men miners' joined or were targeted into the Royal Army Medical Corps (RAMC).

The second safety innovation that proved to be of immense importance when adapted during the Great War, was the establishment of mines rescue stations. Beginning at Tankersley, near Barnsley in 1902, the new stations spread to cover major coalfield areas: south Wales (Abercam), Scotland (Cowdenbeath), County Durham (Elswick), Lancashire (Howe Bridge) and Nottinghamshire (Mansfield). About 6,500 miners were already trained

Mines rescue men with their numbered Proto-branded SCBAs and associated equipment outside the Doncaster station, established in 1913. Brian Elliott

and familiar with mine rescue work at the start of the war.¹² By 1918, there were forty-six strategically-placed Central Rescue Stations and larger collieries had to have a permanent corps of trained brigades on site, under the leadership of a supervisor. In these new safety hubs men were trained in the use of a variety of equipment, but especially self-contained breathing apparatus (SCBAs). The transference of expertise and equipment of the mines rescue service to the Western Front was to be of vital importance in the 'underground war'.

Miners were well known for their bravery during accidents and disasters. The introduction of the Edward Medal in 1907 meant that there was now a national gallantry award. The Royal Warrant for the new medal included the phrase 'to mark the many heroic acts performed by miners and quarrymen and others who endanger their own lives in saving or endeavouring to save the lives of others . . .'. It was almost entirely issued to miners (apart from a small number of service personnel involved in explosive factory incidents) during the War, and became commonly known as the 'miner's VC'. Two classes of award were created: First (silver) and Second (bronze). In the four-year period 1910-13 a total of 131 Edward Medals were presented, including thirteen of First Class standard. This number was 'inflated' because of the sixty-six medals issued for bravery in a single mine disaster, at Wellington Pit, Whitehaven. Fifty-five Edward Medals were approved for British miners during the First World War, ten of the recipients getting First Class awards.¹³

By January 1917, after conscription had been running for a year, and when most mineworkers were largely 'exempted', recruitment still continued, increasing the total of 'miner-soldiers' to 282,200. This was largely due to the great demand for skilled men, for use in tunnelling and other specialist areas of the military services. Writing in 1923, Sir R.A.S. Redmayne, the former Chief Inspector of Mines and wartime adviser to the Coal Controller, estimated that the aggregate number of miners who served in what was now called the Great War was about half a million, an astonishing figure from a single industry.¹⁴

The design on the reverse of the Edward Medal depicts the first stage of the rescue of an injured miner, with the inscription 'COURAGE' shown in the background.

Colliery Guardian, 21 February 1908

King Coal • 17

Chapter Two

1914:
PIT DUDS TO KHAKI

From no class of the community did this call on their patriotism meet with more spontaneous and conspicuous answer than from coal-miners.

(Richard Redmayne, 1923)[1]

Knighted for his services to the coal industry in 1914, Redmayne served as the first Chief Inspector of Mines throughout the war years, and from 1917 was the main advisor to the Coal Controller. But was the great initial rush to the recruiting offices in pit towns due to high ideals? In reality, community and fraternal factors had as strong a pull as King and Country. Living in close-knit pit villages, and working in an environment where companionship and camaraderie was so important, it is not surprising that joining up with workmates was such a popular choice. The miners were perfect for the Pals battalions where working-class men from similar occupations travelled to recruiting offices side by side.

For many young miners, to be left out of the call to arms was unthinkable, especially if they were single or had reassurances that their families would be supported, and that their jobs would be waiting for them when they returned. What was there to lose? And Kitchener's New Army provided an escape route from one of the most dangerous and dirty of jobs imaginable, and one where regular wages were interrupted by stoppages and strikes. In many instances a wife and family was financially better off *after* the enlistment of the main bread winner.

Duty was also tempered by a sense of adventure, or 'an outing on the Continent', maybe.[2] Picture-goers in pit village cinemas could now mimic the heroics portrayed on the big screen; and newspaper reports of the desperate straits of the regular soldiers in France fuelled the enthusiasm to enlist. After all, the war would be over in a few months and there would be a glorious return to home and work.

One motivational factor, however, is never mentioned in military books. Many thousands of miners worked on a sub-contractual basis, in small teams, paid by a leader who shared out the cash in the pit or pub yard. Disenchantment occurred where one man was 'not pulling his weight', so some men were so annoyed with the 'butty' system, that joining up was a welcome relief to receive fair and organised pay.

Published appeals in the press enhanced rather than drove the rush to arms. Posters may have played a part though care is needed in interpreting the impact of Alfred Leete's 'pointing figure' poster of Secretary of State for War Kitchener, issued initially in a London society magazine in September 1914.[3] In any case most miners were not fooled by propagandist posters; they responded more to handwritten, pinned-up notices at pitheads – and chats with mates.

Another underrated motivational factor about 'volunteering' concerned the long family tradition of service in the military. At most pits there were veterans with overseas service in South Africa, India, Burma and China. Pit ponies and miners' children were named after Boer War heroes or events. An unknown number of former miners were already in service as Regulars, their presence bound to be an influence within their own families, friends and communities; and Reservists, back in the mines, alongside part-time Territorials, were the first to report for

Lord Kitchener, seen here on a patriotic postcard. Brian Elliott

A Boer War veteran miner enlists again

WALTER DYSON (1876-1958) enlisted in the York and Lancasters in 1894, serving in southern Africa during the Matabele campaign, and then the South Africa (Second Boer) War. He survived the shipwreck of the RMIS troop ship *Warren Hastings* on 14 January 1897, leaving the army in 1906. Working at Rotherham Main, married with two young children, he rejoined his old regiment (as a Special Reservist) at the age of thirty-eight on 31 August 1914 and was in France by 19 May 1915. Shot in the forearm, Walter was awarded the Silver War Badge and discharged on 24 December 1917, the culmination of an extended military service.[4] There were 'Walter Dysons' in every mining village.

duty, many of them dispatched to France as part of the BEF. These men were often the hewers and haulage hands who were key workers in their pits. The smaller the pit, the more they were missed.

Thus much of the 'summer and autumn madness' in 1914 saw a not inconsiderable migration of experienced pitmen from the mines of Britain, a situation that was bound to encourage workmates to come forward later. After a night at the miners' institute, working men's club or pub, the prospect of 'doing a bit of digging' as part of 'Kitchener's Mob' was perceived as child's play compared with mining coal in low, wet seams.

The rush to the recruiting offices

Local and regional newspapers in coalfield regions provide a wealth of information about miners and the war. The weekly trade journal, the *Colliery Guardian* was also at the forefront of reporting general and regional news relating to the war and the industry.[5]

The *Colliery Guardian*'s editorial thought that the patriotism of the miners was far from universal at the start of hostilities, especially in several of the big, politically more powerful coalfields where there was frequent unrest over wages and conditions of work. In south Wales, it was reported that miners' leaders were urging their members *not to break their holidays* in order to 'hew coal for the Admiralty'.[6] The 'steam coal', the anthracite mined in the Valleys, was indeed essential fuel for the Navy's warships, so the produce of the Welsh miners was a vital strategic resource. The attitude of the president of the South Wales Miners' Federation, William Brace, was – very unfairly – singled out by the *Colliery Guardian* for his 'anti-war' stance.[7] However, by late August, Brace was one of the most powerful voices at recruiting offices in support of Kitchener's call to arms. His stance reflected the unions' change of policy about the War. However, 'pacifism' certainly dimmed after the battle of Mons. It was on 23 August that the BEF had fought its first serious action in the coalfields around Mons, a somewhat ironic setting for the miner-Reservists deployed there.

In south Wales, the response was massive, about 30,000 of the total miners employed in July 1914 had 'drained away' to the armed forces within a few months, mostly young men between aged between eighteen and forty. In the Aberdare valley, 3,000 had 'joined the Colours, practically all [of them] miners'; and about 800 miners had joined the Rhymney Valley Battalion of the Welsh Regiment.[8] One 'young collier' from the Rhondda, rejected as he was below the minimum height, adjourned to a pub where his mates subjected him to 'rack [stretching] drill', 'pulling at his head and shoulders'! For men joining the miners' battalions, the War Office reduced the height of recruits by three inches, to 5 feet 3 inches 'as the South Wales miner is a short type of man'.[9] Under the headline 'Miners in Arms', the *Western Mail* reported the spectacle of 500-600 volunteers of the 10th (Rhondda) Battalion of the Welsh

Punch magazine, not read by many miners, published this recruiting scene in its 25 November 1915 edition. *Punch*/Brian Elliott

Regiment, parading in Sophia Gardens, Cardiff.[10] In an extraordinary turn around, the War Office appealed to miners *not to enlist* in the South Wales Borderers 'as it was becoming very difficult to work the mines'.[11]

> **MEN OF DURHAM.**
>
> LORD KITCHENER, Secretary for War, invites our co-operation in the work of raising the additional number of regular troops required at once for the Army.
>
> 100,000 MEN are immediately required; I appeal to you with the utmost confidence to uphold the old traditions of the County and the glorious records of our Faithful Durhams by showing in the present grave emergency that our patriotism impels us to make every sacrifice and every personal effort in defence of the honour of our Empire.
>
> LET DURHAM NOT LAG BEHIND in the formation of this second army, but prove that we intend to be amongst the foremost in filling its ranks.
>
> READ LORD KITCHENER'S APPEAL, offer yourselves in thousands as recruits, firstly for the Regular Army, secondly for our Territorials.
>
> Act like the brave men you are, and Durham will nobly respond to the Nation's call to duty.
>
> DURHAM,
> Lord Lieutenant,
> and President Territorial Force
> Association of the County of Durham.
>
> **GOD SAVE THE KING.**

Many Durham miners read this appeal, printed in the *Durham Chronicle* shortly after war was declared, 14 August 1914.

Durham Chronicle/Brian Elliott

In Scotland, hard-pressed clerks at Kirkcaldy, Fife, struggled to cope with 500 recruits a week, about 75% of them miners.[12] A month later, about 4,500 members of the Lanarkshire Miners' Union had enlisted;[13] and recruitment for the Army or the Navy had reached 1,850 from the colliery districts that made up the Mid and East Lothian Miners' Association.

In the old mining areas of Northumberland and County Durham, miners came forward in 'alarming numbers'. Within days of Kitchener's appeal, a 'special telegram' from Newcastle stated that 'large numbers of men are changing their pit duds for khaki' and that there was 'a constant stream of volunteers at the recruiting offices ... in the mining villages'.[14] Individual pits where returns had been completed showed that 'losses to the military' were considerable: Ashington, 800 men gone; Bedlington, 500; Walker, 200; New Delaval, 110; Seaton Delaval, 100 and Hartley, 100; contributing to an estimated exit of 12,000 North East miners in a few weeks. When Lieutenant F.R.G. Shiel, the former manager of South Garesfield Colliery, appealed for fifty men from his former pit to complete his battalion in the Royal Field Artillery his target was fulfilled in a day.[15] By 20 October, it was estimated that 20,000 men from the Durham Miners' Association had volunteered.[16] County Durham's Blackhall Colliery was forced to close down 'as over a thousand of the employees have joined the Regular Army or the Territorials'.[17]

A somewhat unusual request occurred at a Manchester district recruiting station when three Irish miners demanded to be 'real soldiers' and not 'these toffy soldiers of Kitchener's'. Along with another pal, they were placed in 'Paddy O'Rourke's' Irish Fusiliers.[18] But generally, recruitment from the Lancashire coalfield matched the response elsewhere. Of the 4,049 employees of the Lancashire colliery owners Andrew Knowles and Sons Ltd, 585 (14%) had 'joined the Colours'.[19]

Newspapers from the geographically huge Yorkshire, Nottinghamshire and Derbyshire coalfield reported on the large numbers of miners enlisting during August. In Mansfield, the town hall could not cope with almost 400 volunteers a day.[20] The Pontefract-based West Yorkshire Coal Owners' Association planned to raise and equip a special battalion of miners (12th [Pioneers] King's Own Yorkshire Light Infantry [KOYLI]).[21] The same report also stated that the new large pits in the Doncaster area were at the forefront of recruitment, the 5th Battalion KOYLI said to be working with Viscount Hickleton and 'other influential gentlemen' in appeals; but the prospect of a separate Doncaster miners' battalion had 'fallen flat'. In addition, 'a splendid battalion' of the 5th York and Lancasters was now almost fully enrolled in Barnsley, 'the backbone supplied by miners'.[22] The first (13th Service) 'Barnsley Pals' battalion – and the formation of a second – was well underway. Lieutenant-Colonel Sir Joseph Hewitt, the new commander of the 'Barnsley Battalion', was quoted as saying that Yorkshire miners had already contributed 16,000 men 'to the Colours'.[23]

Lieutenant-Colonel Sir Joseph Hewitt, the commander of the new 'Barnsley Battalion', was managing director of the colliery company Fountain & Burnley Ltd and president of the Barnsley and District Colliery Owners' Association. *Colliery Guardian*/Pen & Sword

Support for miners and their families

Although there was no legal requirement, many colliery companies provided support for their workers who had either volunteered or had been recalled to military service. The owners of St John's Colliery, Wakefield, gave £3 to each its workmen who had enlisted, and 5s a week was to be set aside until employment was resumed.[24] In south Lancashire, the Clifton and Kersley Coal Company authorised allowances of 5s per week for a wife and 6d for each child.[25] The owners of Thrislington Colliery, West Cornforth, County Durham gave 2s to each volunteer's wife, and an additional 1s for every child. John Bowes and Partners' Collieries also had a similar scale of allowances for the families of men serving abroad, a wife and two children getting 6s a week.[26] Such payments were fairly typical of those given in other areas and were in addition to the state 'separation allowances' or 'war pay' provided via the Post Office. Some coal companies even waived the payment of rents in their houses occupied by wives and children of recruited miners; and residents were not charged for home coal. In south Wales, the huge Powell Duffryn and Bute company gave 10s a week (and 1s for each child) to 'enlisted families',[27] though 'better allowances for families and dependants' were required, according to a report from the Rhondda.[28]

There were many instance of the miners themselves – and their unions – making contributions towards the assistance of families whose loved ones had left for the military. At Brodsworth Main, Doncaster, one of the largest and most profitable collieries in the country, the workmen voluntarily contributed 1d a shift per man to a relief fund, amounting to about £43 each week. The committee responsible for administering and distributing payments allocated 1s 4d to every 'child-occupied' household, more than doubling the government's separation allowance, and benefiting 200 children.[29]

Anti-German feeling

Despite the War, the sinking of exploratory boreholes and shafts in the concealed coalfields continued, though not without difficulties due to public reaction to the involvement of German industrialists, engineers and labour. 'Anti-Hun' feeling also surfaced at several of the new collieries around Doncaster. A plan to start preparation work at Markham Main was suspended; and, further eastwards, work actually ceased at Thorne Colliery, the German shaft-freezing or cementation process used in pit sinking being totally abandoned.[30] At Harworth (north Nottinghamshire), where German capital was heavily invested, the colliery company's assets were impounded, sinking suspended and the German workers interned. The Aliens Restriction Act (5 August 1914) meant that German or Austrian people living in Britain were subjected to compulsory registration, and categorized as 'enemy aliens'. Any 'Austro-Germans' assessed as security risks were 'detained'.[31] From as early as September 1914 Lofthouse Park, near Wakefield, was used as a prison camp for 'better off' Germans.[32]

In the new Kent coalfield 'military authorities' were concerned about the strategic knowledge and know-how of 'foreigners' as German workers had been employed in boring and shaft-sinking operations.[33] Questions were asked in Parliament about the sinking of Chislet Colliery because of its German connections, clearly evident in the name of its parent company, Anglo-Westphalian Coal Syndicates Ltd. It quickly metamorphosed into 'North Kent Coalfield Ltd'!

Belgian refugees

Displaced families from Antwerp and Louvain – 'Belgian refugees' – were made welcome in coalfield towns and villages. Thirty-seven had arrived at Denaby Main via Conisbrough station, in south Yorkshire, in October. The Denaby and Cadeby colliery company housed them rent-free in ten of their houses, and the miners themselves subscribed £6 a week towards refugees' expenses.[34] In Barnsley, an 'enthusiastic reception' was given to forty-one Belgian refugees, the number increasing to sixty-four by the end of October.[35] Towards the end of the year, about 750 Belgian refugees were housed in sixty-

Belgian refugees in Barnsley, photographed in 1914 for a postcard. Chris Sharp

one Derbyshire parishes.[36] Few of the incomers had mining backgrounds but in late December Belgians were actually employed in collieries in the Manchester and Bolton districts, where the enlistment of colliers had caused a serious shortage of labour.[37]

'War reports' for 1914

The divisional mines inspectors' reports confirm the great exodus of miners from the pits to the military during the five months after war was declared. The Northern Division (Northumberland, Durham and Cumberland) lost 46,416 of its workforce, or 18.7%. The South Wales Division was depleted by 30,127, 12.9% of its labour; and employment in the Yorkshire and the Midlands mines was reduced by 9.3%, or 25,674 men.

The massive reduction in the export of coal was also mentioned by the Northern Division's inspector, who said that many of the coastal pits had closed down *for the year*, and miners at other 'exporting pits' were on short-time work. J.R. Wilson's report was the most detailed one relating to the war, and he duly paid tribute to the response of 'his' miners:

> Since the British Empire became involved in War, the miners of the Northern counties have responded to the Call to Arms in a manner

which has kindled very pardonable local pride; and the large numbers which have joined the forces and the enthusiasm which have been displayed in the work of training will be remembered with very keen satisfaction for many a year to come.

Wilson now noted that '49,000 men . . . drawn from coal mines . . .' had left his Division 'and become soldiers'. Some of his collieries had an exceptional response. At Ryhope, 874 persons (from a workforce of 2,573 [34%]) had 'joined the Forces', a proportion much higher in reality (63.75%) since only 1,370 of the total employees 'were of military age'(19 to 38 years). At Whitehaven collieries, 691 enlisted from 1,296 employed, over 53%. Wilson praised the large Ashington group of collieries in Northumberland where 2,517 (from 9,414, 27%) had responded. This cohort included 1,800 trained ambulance men '[whose] knowledge of First Aid possessed by the latter must have been of great use to their comrades who have been wounded in the field . . .'.

Thomas Mottram, the Yorkshire and North Midlands Division inspector, concluded that the war had reduced employment at *every colliery*, the proportion of loss ranging from 20 to 40%. Replacements, he said, were largely of boys aged over fourteen years of age, but most of the vacancies remained unfilled.

The Lancashire and North Wales inspector, A.D. Nicholson, noted that 'recruiting had been particularly brisk among the colliers around St Helens, Wigan and Manchester districts', the proportion of 'leavers' consisting mostly between 10 and 30% of workforces. The greatest 'escape from the dark' came from St Helens where 'about 85% of those of military age' had enlisted.

The mines inspectorate itself was also disadvantaged by the war. A Scottish inspector, G.K. Robinson, was in France, commanding a battalion of the Monmouthshire Regiment. Three sub-inspectors from the Northern and Yorkshire and Midland divisions had volunteered 'for duty'; and four staff were 'absent to the military' from the South Wales Division.

Two 'miner-poets'

The miners who enlisted or were recalled in 1914 were mostly literate, able to express their feelings in writing, and many of course did so quite freely, with little or no censorship to hamper their efforts. Abstracts and whole letters 'from the Front' began to appear in local newspapers on a regular basis, with poems popular insertions.

Under the heading 'Poetry from the Trenches', **Corporal William Brodie**, from the New Stevenston and Holytown area of Lanarkshire – 'Keir Hardie country' – expressed his feelings via a poem in the

Corporal William Brodie.
Brian Elliott

Christmas Day edition of the *Motherwell Times*. The last verse (of three) of Brodie's patriotic message was as follows:

> Oh Scotsmen, brither Scotsmen,
> If you could see things here.
> Your heart would bleed for Belgium,
> And your eyes would gush with tears;
> 'Then up and at Them,' Scotsmen,
> As your fathers did of yore,
> Until poor Belgium's freedom
> And France's you restore.

No Owen or Sassoon, but Brodie's heartfelt lines were read by tens of thousands of readers; so much so that a leaflet was printed, widening the interest further, 'the proceeds devoted to patriotic purposes'. A former Blackie Colliery miner, Brodie, aged twenty-seven, served with the Argyle and Sutherland Highlanders, and was 'wounded in action' a few weeks prior to the penning of his 'trench poem'. Sadly, his war came to a sudden end when he died in a field ambulance on 29 March 1915, after hospitalisation for septicaemia. He was interred in France at the Cite Bonjean Military Cemetery, Armentières.

On 15 September 1914, **John William ('Will') Streets**, a young Wesleyan Sunday school teacher from Whitwell, Derbyshire, was medically examined for the York and Lancasters, in Sheffield, and passed fit to serve in the 12th (Service) or Sheffield City Battalion, popularly known as the 'Sheffield Pals'. Soldier number 525, Streets, was twenty-eight, the youngest in a family of twelve children. From age fourteen he had worked as a miner at Whitwell Colliery, at first pony-driving and then hewing coal.

Sergeant John William Streets.
A. St. John Adcock, For Remembrance. Soldier Poets who have Fallen in the War (Hodder & Stoughton, 1920), p.192/Public Domain

Will spent fifteen months training with his battalion, prior to a two-month period of overseas service in Alexandria, guarding and defending Suez. What followed, from a base opposite the German front line, at Serre, France, was a precarious presence amid one of the most notorious killing fields of the First World War. Promoted to Sergeant, on 1 July, from behind John Copse, Will was part of a second wave of Sheffield Pals that advanced, laying in wait in No Man's Land. Attempting an assault on the German front line, they were met – according to the battalion War Diary – with 'very heavy machine gun and rifle fire and artillery barrage'. It was a hopeless situation, with few soldiers surviving intact and not one breaching the German wire. Wounded, Will managed to get back towards a First Aid point but went, unselfishly, to help another casualty. Never seen alive again, William Streets was one of 468 of the other ranks of Sheffield Pals killed, wounded or missing. His body lay undiscovered for ten months, until identification on 1 May 1917. Burial was at Euston Road Cemetery, Colincamps.

Two pocket books listed as part of his effects contain many of Streets' compositions, several of them already sent home to his parents at Whitwell. The poems were, in Will's own words 'inspired while I was in the trenches' of the Somme. *The Undying Splendour* was posthumously published in a little volume by Erskine Macdonald, in May 1917.[38] The compositions included several written shortly before his death: *A Lark Above the Trenches*, *Shelly in the Trenches*, *April Evening: France, 1916* and *Matthew Copse*. In *The Call*, part of an opening sonnet sequence, Streets was at his most patriotic, concluding with the lines

> And lion Youth fearless and rampart came
> To follow Freedom's flag into strife.
> Old England saw proud Youth allegiance take –
> Men with souls of Wellington and Drake.

This former Derbyshire miner who lost his life at the age of thirty-one on the first day of the Battle of the Somme, deserves more recognition as a notable Great War poet.[39]

'Miner-birds' for the Army

Pigeons, for long part of the miners' cultural life, were co-opted by the War Office for their unique carrier capabilities, attributes that were felt to be far less vulnerable to enemy 'tapping' than wireless messages. Over short, land-based distances and indeed over the sea to Belgium, the most notable 'pigeon country' in Europe, there was nothing better or quicker to deliver a message than 'a fast bird'. Back in Blighty, miner-pigeon fanciers such as Fred Elliott, my paternal grandfather, had to have permits to retain their stock of birds.[40]

Two miner-authors

The writings of **'Frank Richards'** (Francis Philip Woodruff, 1883-1936), provide us with a remarkable account of one miner's rapid return to arms.[41] Aged thirty-one, a Reservist who had served with the Royal Welch from 1901 to 1909, Richards had returned to pitwork, as a 'timberman' (setting roof supports). On 4 August, on the very day that Britain declared war on Germany, he met with a small group of ex-soldier-miners in the Castle Hotel, Blaina, south Wales. Ale-infused reminiscences dominated conversation and within a few days he was serving in France.

Richards had an eventful and fortuitous war, surviving action in most of the major battles on the

Frank Richards.
Margaret Holmes/Glo[Coal]: Big Pit NMW

Western Front. Here is part of his account of Christmas Day 1914, a respite from normal front-line duty:

> We and the Germans met in the middle of no-man's-land. Their officers were also out now. Our officers exchanged greetings with them . . . We mucked in all day with one another. They were Saxons and some of them could speak English. By the look of them their trenches were as bad as state as our own . . . The officers came to an understanding that the unofficial truce would end at midnight. At dusk we went back to our respective trenches.[42]

Richards remained a private but was regarded as the 'best signalman in the regiment' by one of his most notable officers, the soon-to be-famous author, Robert Graves. Another literary commander, Siegfried Sassoon, also held Frank in very high regard.

Frank Richards went on to win the DCM in February 1916, in an attack on a German-held crater. Following further heroics during an attack on Polygon Wood (3rd Battle of Ypres [Passchendaele]) in 1917, he got yet another gallantry award, the Military Medal. It was indeed an extraordinary war for the veteran Welsh miner.

Richards' military service ended in 1919, on discharge. His home life, however, was troubled by debilitating rheumatism, due to spending so much time in the mud and water of the trenches. Unable to find work in the mines, he received little in the way of support from both the colliery owners and the military. Disheartened and disillusioned, he was understandably bitter about his treatment but managed to put down his wartime experiences in writing. The outcome was what became the classic memoir *Old Soldiers Never Die* (1933), one of the best first-hand accounts of a rank-and-file soldier's life in the Great War.

For most miner-enlistees joining the Colours meant army service, but a significant number preferred a 'life on the ocean wave', especially when there was a family connection with the Royal Navy. **Joseph ('Joe') Murray's** choice was influenced by the self-confessed hero worship of his older brother, Tom, in the Navy from 1906.

The son of a Scottish miner and former Northumberland Fusilier, Joe lived in the one-street pit village of Lintz, Burnopfield, in County Durham. In this reminiscence he describes his trip to enlist: 'On Tuesday, 6th October, after a hard day's work, I walked the eight miles to Elswick and volunteered to join the Royal Navy Volunteer Reserve (RNVR) on board HMS *Calliope* on the Tyne.'[43]

'Work' for Joe was an eight to ten-hour stint at South Garesfield Colliery where he had toiled from

Joseph Murray aka 'Lucky Durham' in 1915.
Call to Arms, Kimber & Co, 1980

Good Hope (left) and *Monmouth* (right) were both sunk by the German navy off Coronel, central Chile, in 1914. Taylor Library

the age of twelve, for 10d (4.75p) a shift. It wasn't long before he was on his way to Gallipoli. Relatively unscathed, 'Lucky Durham' – unlike so many of his miner mates – was soon on the Western Front (attached to the Hood Battalion of the Royal Naval Division), remaining there until sustaining a wound at Gravrelle in 1917.

Many years later, Murray was able to recount his very active military experiences in a most vivid way, based on his own 'in the field' notes. *Gallipoli as I saw it* (1965) and *Call to Arms* (1980) remain compelling first-hand accounts of the disastrous Dardanelles campaign. Throughout his life, Joseph always had thoughts in mind about older brother Tom, who lost his life when his ship the *Good Hope* was sunk with *Monmouth* in a battle with the German navy off Coronel, central Chile, on 1 November 1914. The 1,600 officers and men who lost their lives from both ships included numerous former miners.

Royal Army Medical Corps (RAMC)

Many miners were welcomed into the RAMC because of their qualifications and abilities in first aid and mine rescue. Although still quite young, such 'candidates' had often passed one or more of the quinquennial tests offered by the regional St John's Ambulance services, had taken part in First Aid competitions and undergone rescue and resuscitation training. Such 'ambulance men' were of immense value to the army, therefore fast-tracked into overseas service with the RAMC, a non-combat force, so recruits needed little or no military training. And yet over time many faced as much action than most infantrymen.

Twenty-four young miners (aged 22-36) volunteered for the RAMC from one small mining town, Mexborough, in south Yorkshire, within days of the start of the war.[44] These welcome recruits helped to boost the existing c.5,000 cohort of the RAMC to around 9,000 by the end of August 1914, a figure that rocketed to 155,000 (other ranks only) by 1918. What a 'hell on earth' baptism the initial ones had during the retreat from Mons, overwhelmed with so many casualties, the medical services nearing total collapse. Later, they were subjected to the horrors of front-line war, 4,139 RAMC being killed in action or dying later.[45]

Tribute medal presented after the war to Harold Fitch by the Wath Main Coal Company. Brian Elliott

Harold Fitch, aged twenty-three, a Wath Main Rescue Station man, and one of twelve members of the Denaby Main Ambulance Corps that made up the Mexborough volunteers, was in France on 21 August 1914. He was allocated to Number 4 Field Ambulance (FA), a unit consisting of about ten officers and 220 other ranks. The 4/FA were kept extremely busy during the Mons debacle, carrying, conveying and dealing with the wounded and the dying between Aid Posts and Casualty Clearing Stations. Harold's frontline medical activities were severely curtailed on 28 September 1915, during the Battle of Loos, when he was shot in the left arm and poisoned by gas. Discharged only a few weeks later, he was then detailed to work on a hospital ship, his last active posting, until transfer to the Army Reserve Class 'P' (enabling him to resume work as a miner).[46]

Studio photograph of Private Harold Fitch, France c.1914. Brian Elliott

Aged twenty-eight, Private Harold Fitch got his army discharge notice in January 1919. Apart from receiving the British War and Victory medals, he was also entitled to the Silver War Badge on account of the wounds and respiratory ailments obtained earlier. Coping with poison gas-inflicted chronic bronchitis, and goodness knows how many mental scars from his war experiences, employment in mining, at least underground, was probably limited, though he may have carried on with his ambulance work. Fitch was one of about a hundred pitmen presented with Tribute medals from the Wath Main Coal Company after the war.[47]

1914: Pit Duds to Khaki

Two Barnsley pals sing *It's a long way to Tipperary*

Still in his 'pit muck' after a shift of pony-driving, **Clifford ('Cliff') Tyas**, aged nineteen, sat in the back room of 13 Fitzwilliam Street, on Friday afternoon, 26 September 1914, reading the front page of the *Barnsley Chronicle* with more than passing interest. The 'SPLENDID RECRUITING PROGRESS' of the fledgling 13th battalion (Barnsley) of the York and Lancasters was headline news, thanks to the efforts of the leading enthusiasts, most notably a local solicitor, Joseph Hewitt (its acting colonel and commanding officer), and the mayor, Alderman William Elmsley Raley (Hewitt's deputy). Among the initiatives were speeches that Cliff may well have heard live rather than read in print, for example the half-time 'rallying' during a dull goalless draw at the Barnsley versus Grimsby football match the previous Saturday afternoon.[48]

Clifford Tyas, Great War veteran, photographed at his Golden Wedding in 1969. *Barnsley Chronicle*

Also reported was a large open-air meeting attended by 2000 miners, including several hundred bleary-eyed and black-faced day-shift men. Raley's speech was perfect for the occasion, referring to the miners as 'the cream of the [New] Army', workers who 'faced death every day'. Some 200 miners – as if at a religious assembly – 'came forward', putting their names down at the colliery offices, adding to 113 of their colleagues already enlisted. The day after, the volunteers marched to Barnsley centre, singing *It's a long way to Tipperary*.

Distinguished Conduct Medal. Brian Elliott

On Monday, 28 September, now wearing his 'better clothes' including a floppy flat cap, Cliff walked the short distance from his home and joined the queue that began to spill on to the pavements outside the Harvey Institute (Public Hall). At 5 feet 5 inches and 117 pounds, his modest physique was not a problem for the recruitment officer.

Assigned number 952 of the near-complete 13th Barnsley (Service) Battalion of the York and Lancasters, Tyas got fitter and stronger from drill exercises and brisk marches with his mining mates. The outdoor exercise was hard and tiring but a lot easier than underground pit work.

Cliff soon saw bits of Britain that he had never seen or heard of before. A few days after Christmas 1914, the 13th Barnsley were marched over four miles to their base at Newhall Camp, in partly cleared woodland above Silkstone village. Manoeuvres, mock combat, trench digging and route marching took place in dour winter conditions, a foretaste of what was to come. Cliff Tyas survived the war, unlike many of his friends who fell on the Somme, and when he was demobbed had a DCM, awarded in 1918 when he was one of the few remaining original Barnsley Pals.

In August 1914, **Gilbert (Gil') Hall**, aged twenty-two, was employed as a surface worker at Elsecar Main Colliery, near Barnsley. His Staffordshire-born father, Samuel, was a blacksmith's striker in the workshops at the same pit. The weather the day after the August Bank Holiday was very warm and sunny, so Gil spent the afternoon (4 August) wandering around the fields behind the village with five mates, when they were told by excited locals that Belgium had been invaded and war was expected. The pals, with little experience beyond their homes, got caught up in the hysteria of the occasion – an opportunity to have a continental holiday at the government's expense.

Rejected at a local recruiting office (as there were 'too many miners'!), the enthusiasm of the six friends was restored following a packed public meeting held in Hoyland Town Hall on 24 September, speakers appealing for recruits to a Pals battalion to be raised in Barnsley; and the next day all six lads attested in the Harvey Institute, only a few hours after Cliff Tyas had 'signed on' there. Reporting for duty, there were no uniforms, no arms or equipment for them but they were ably taken in hand by a couple experienced warrant officers, veterans of the Boer War. Gilbert and his friends' training experiences matched those of Cliff, and their stories remained similar through to the Pals' dispatch to Egypt in early January, in the momentous year of 1916.

Gilbert Hall in 1917, after his commission with 2/4th KOYLI. P. Haigh & P.W. Turner, *Not For Glory*, Maxwell, 1969

Hall's experiences when he was part of A Company on the Somme and were horrific and nightmare-making. Whilst the original 13th Barnsley was dramatically depleted and 'unrecognisable', Gilbert rose above all the odds to get a commission in the 2/4th York and Lancasters (KOYLI's), though a shell splinter piercing his throat ended his active service, a few weeks before the Armistice. Remarkably, he and his enlistee mates survived the war. Gilbert's story was retold over fifty years later in a now largely forgotten book, *Not For Glory*.[49]

A 'one-legged miner'

A Reservist, **Walter ('Walt') Ackroyd** was of a small stature: 5 feet 5 inches and ten stone, when he rejoined the York and Lancasters at the age of twenty-eight at the end of August 1914. Married with two young children, by 1911 he had advanced from haulage jobs at a Rotherham colliery to 'hewer'. By August 1915, he was also the father of a three-month-old baby girl, Margaret, who must have occupied his mind when away from home.

Ackroyd's subsequent military experiences spanned the entire war. He served with the BEF 'in the Mediterranean' (Egypt) and then on the Western Front experienced most of the major battlegrounds, including the Somme. His service was not without harm. A gunshot wound to his right elbow was followed by a bullet entering his left thigh, which in 1917 resulted in amputation, a 'dangerously ill' comment on his file and many weeks in military hospitals. Prior to his demise, Walter had achieved several promotions, culminating in the rank of Sergeant, prior to being discharged from the army as 'no longer physically fit for war', in June 1919.

Astonishingly, with the help of crutches, Walt returned to mining, working as an underground engine driver, not the easiest of jobs for an able-bodied man never mind one badly disabled. What a character he must have been. A widower since 1927, he devoted his spare time to public service, serving on the Rotherham Borough Council. A heart attack sustained whilst at work ended an extraordinary life at the age of fifty-two in 1952. Ackroyd's funeral was attended by many town worthies and officials, as well as union representatives. Hundreds of mourners hailed him as a 'war hero' and the local press took delight in referring to him as the amazing 'one-legged miner'.[50]

Contrasting 'miner-recruits'

From 1900 to 1914, about 2,500 miners had settled in Grimethorpe, attracted by the prospect of work at a large, modern Yorkshire colliery. Amongst the later arrivals was a Lancashire pitman and his wife, **Patrick Ryan** and Mary Ann. Patrick had little time to get used to his new job at 'Grimey', as he enlisted in the 3rd (Reserve) York and Lancaster Regiment, on 2 September. Ryan was only a month short of the then 38-year-old age limit for recruitment, but already had sixteen years' experience as a regular soldier with the Lancashire Fusiliers. Therefore a Special Reservist, Ryan was on the Western Front within a year. His ability in using arms was recognised in May

Private Patrick Ryan's British War Medal. Brian Elliott

1916 when he was transferred to the Machine Gun Corps, founded a few months earlier, in October 1915. Private Ryan's war came to an end on 8 August 1918, a few weeks before the signing of the Armistice, when he was discharged 'as no longer fit for War Service'. He was transferred to the Army Reserve (Class 'P', for miners), which allowed re-employment as a miner.[51]

On 4 August 1914, the day that Britain declared war on Germany, a boy miner from Ormskirk, Lancashire, went along with a

Ryan's 'mining mates' at Grimethorpe Colliery. Brian Elliott

1914: Pit Duds to Khaki • 35

few of his mates to 'report for duty' at their local recruiting office. **Richard ('Dick') Trafford**'s original attestation form survives. The recruitment officer was correct when he wrote 'collier' and 'Rainford' (Rainford Colliery) in response to his 'trade or calling' and, but the record of his age as 'seventeen' was incorrect. Richard was only fifteen. When the youngster's age was questioned by the officer, Dick offered to pop back home for his birth certificate. The bluff worked. He was now a private in the 1/9th Battalion, King's Liverpool Regiment. Trafford was one of about 250,000 'under-agers' who served in the Great War, an unknown number of them from the mines. Dick's story – based on his own words – is described in Richard Van Emden's landmark book *Boy Soldiers of the Great War*.[52]

But why did Dick Trafford rush to join the Colours? It was not so much 'to escape from the pit', but 'to keep up with other men' that appears to have been his prime motivation.[53] Such a *raison d'etre* was an integral part the camaraderie embedded in the soul of every coalminer. Life and work were inseparable companions. When interviewed by Van Emden, Dick described his perceptions at the eve of war so clearly that we have no doubt as to why he volunteered on that warm summer's day:

> About six men from Ormskirk and me used to be on night shift, and this particular night we'd all turned up for work but one chap was late . . . When he arrived he said, 'There's no work tonight, chaps'; the war was going to break out tomorrow and, he said, we had better go and report to the drill hall . . . They were Territorials and had to report there and I followed them to enlist. It wasn't because I actually wanted to go, it was because other men were going and I thought, *Well, I might as well be with them*.[54]

Trafford's teenage military experience included some of the bloodiest battles and scenes of the war, at Loos and on the Somme. And then he survived the mud and death-splattered landscape of Passchendaele, as well as the high-fatality campaigns of 1918, despite being wounded and gassed. In 1919, he returned to Ormskirk, starved and exhausted. Any residual elation was tempered and then lost when he struggled to find work. But he had no regrets about 'his war': '. . . because I knew what I was doing. I must have done it, that's the way I look at it.'[55]

What a shock it must have been for Tom and Jane Spencer when, in June 1917, just before their paternal grandson was due to be on leave, news reached them that their son Private **Leonard ('Len') Spencer** had been killed on the Western Front. They had cared for him as though he was their own son since he was a small boy. News travels fast in close-knit pit communities like Thurnscoe, in south Yorkshire, but the wider public

Private Leonard Spencer. Brian Elliott

Spencer's 1914-15 Star medal. Brian Elliott

soon got to know about Private Spencer via their local newspaper, presented under a sadly all too familiar banner headline: 'Local Men who Have Fallen'.[56] A small grainy picture shows Leonard amongst more than twenty other portrait images, mostly of young former miners who had been wounded or killed in action.

Although Spencer's service papers have not survived, the short account of his life, in a column running alongside the images, provides useful albeit not entirely confirmatory information about some of his background. He worked underground

The war memorial in Thurnscoe Park, unveiled in 1920, was made at a cost of £750, funded by the Hickleton Main Coal Company and its workmen. Spencer was one of 82 casualties inscribed on the original memorial plaques relating to the First World War (and three more added to later plaques). Brian Elliott

1914: Pit Duds to Khaki

at Hickleton Main, maybe alongside his grandad, enlisting with the 8th (Service) Battalion York and Lancasters in August 1914. The regimental War Diary shows that a few days before his death, on 29 May 1917, the 8th had relieved the 10th West Riding Regiment at Hill 60, prior to trench duty near Zillebeke, when they encountered heavy shelling 'in retaliation for our own continued artillery service'. It was probably during this German onslaught that Len lost his life, at the age of twenty-one.

Leonard Spencer was entitled to the 1914/15 Star medal as well as the usual War and Victory medals.[57] He was interred in the Hop Store Cemetery, Ypres. His grandparents may have got some solace when they received his Memorial Plaque, also known as 'death plaque' or 'dead man's penny', and perhaps even more so when his name was inscribed on the village war memorial.

When, in the autumn of 1914, the Leeds-based West Yorkshire Coal Owners' Association began forming a miners' battalion (12th [Service] [Miners'] Battalion, King's Own Light Infantry [KOYLI]), a **Ernest ('Ernie') Hullock**, a teenage collier from Hunslet was one of the first to come forward, allocated with the roll number 80. It was a pit-head notice that encouraged Ernie to volunteer, according to his attestation form.[58] Hullock's age, correctly noted, was nineteen years and two months and his diminutive physique, at 5

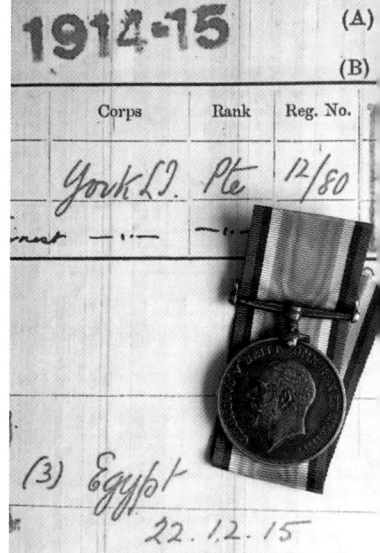

Ernest Hullock's British War Medal, with its low number, '12. 80'. Hullock was also entitled to the 1914-15 Star and Victory medals. Brian Elliott

38 • MINERS AND THE GREAT WAR

feet 4.5 inches and 131 pounds, proved no barrier to enlistment. As a young hewer, however, he had a big asset: strength and fitness, the very hallmarks of so many miners.

Ernest had joined one of the most notable and specialist of the Great War Pals battalions. After training in Yorkshire, and then Salisbury, he and the 12th embarked for Egypt on 6 December 1915. Almost a year later he was discharged from the army. But what a year it had been. On 2 March 1916 he was in France, soon involved with his miner-mates digging deep trenches and dugouts near Serre, as part of the Somme summer offensive. Ernie found such tasks second nature because of his experiences using a pick and shovel; and laying narrow gauge railway lines was routine haulage work as far as he was concerned. Integral to his experience, too, was an awareness and knowledge of the danger of gas emissions, particularly during the deployment of explosives. On 1 July, with a pick and shovel as part of his pack, he was with one of the platoons that took part in an infantry attack, in the face of overwhelming artillery and machine gun fire. Surviving relatively unscathed for a further three months, until 2 October 1916, he was released from the 12th and allowed to return to his job at Beeston Colliery.

Placed on the Army Reserve, Ernest married his sweetheart, Lavinia Beevers, in May 1917. Like so many others, it is not known how the war, and especially the horrors of 1 July and its aftermath, affected his physical and mental state, but he found work as a council labourer, a far lighter job than mining but with far less pay, prior to his somewhat premature death at the age of fifty-eight, in 1953.

The first 'miner-VCs'

During the autumn and early winter months, news reached the coalfields about former miners whose actions in the war were so exceptional that they were awarded the ultimate award for valour, the Victoria Cross (VC). The first of these concerned a 28-year-old Scot, **Private George Wilson**, who had worked at a Niddrie and Behar Company pit, near Edinburgh. On 14 September, near Verneul, France, in a lone 'suicidal' attack on an hostile machine gun placement, Wilson shot six German soldiers and bayoneted their officer. Wounded and gassed at Loos in 1915, and discharged from the army as medically unfit, he died in sad circumstances eleven years later.

Two weeks after Wilson's heroics, on 28 September, at Chavonne, France, a 27-year-old private in the 2nd Battalion, Coldstream Guards, **Frederick William Dobson**, formerly a horse-keeper at Backworth Colliery, Northumberland, won the Victoria Cross for his actions during the Battle of the Aisne. Under heavy enemy fire, he crawled and assisted several wounded comrades stranded on open ground. Hospitalised prior to discharge from the Army in 1917, Dobson's return to mining was limited due to pain from war wounds, passing away at the relatively young age of forty-five.

Another North East man, **Henry ('Harry') Howie Robson**, from South Shields, worked underground as a 'rope lad' (haulage hand), probably in St Hilda's Colliery, prior to joining the Royal Scots in 1912. Robson arrived in France with the Scots on 11 August 1914, saw service at Mons and experienced the battles of Marne and Aisne. After reaching Kemmel, Belgium, on 14 December, during an attack on a German position at Petit Boise, Private Robson left his trench under heavy fire and rescued a wounded officer. He then tried to carry another man to safety, despite getting wounded at the outset, continuing his efforts, even though sustaining a further gunshot injury. Howie was only twenty-one when his VC was finally presented to him at Buckingham Palace, on 12 July 1915. After a long convalescence he returned to the Western Front but on 13 November 1915, at Serre, he was wounded yet again. Harry Robson never returned to mining, selling his VC to pay for his passage to Canada, where he died in 1964.

Robson was one of many VCs celebrated by Gallaher on their Victoria Cross Heroes cigarette cards. Brian Elliott

A few days after Robson's actions, on 21 December 1914 at Rouge-Bancs, near Neuve Chapelle, **Private Abraham Acton** and Private James A. Smith (aka James Alexander Glenn) from the Border Regiment showed 'conspicuous bravery' for voluntarily leaving their trench to rescue a wounded man left exposed by the German trenches for many hours; and then, under heavy fire for an hour, brought to safety another wounded comrade. Acton, a Whitehaven man who had worked with his father at Harrington (Lowca) No 10 Colliery spent a short period as a Territorial before becoming a regular with the 2nd Battalion of the Borderers in January 1914, leaving for France as part of a large draft of BEF reinforcements on 25 November. Sadly, he was unable to receive his VC as he was killed during the Battle of Festubert, on 16 May 1915. Acton's image was used for 'Zam-Buk' (herbal balm) adverts and featured in the first series of 'Victoria Cross Heroes' cards issued by the cigarette makers Gallaher; but he was portrayed more accurately, albeit posthumously, in an oil painting now held in Whitehaven's Beacon Museum.[59]

This painting of Private Abraham Acton by John Dalzell Kenworthy was completed posthumously in 1916 and forms part of the collection at The Beacon Museum at Whitehaven. Beacon Museum

Chapter Three

1915: INNOCENCE LOST

By the end of 1915, any 'innocence' concerning the War had long evaporated.[1] Nowhere was this more evident than among the thousands of miners who had enlisted with such high, optimistic ideals, expecting to be home by Christmas 1914.

News from all fronts was not good as the festive period passed and 1915 dawned. Coalfield newspapers began to publish long lists of casualties, many of them former miners. The *Barnsley Chronicle* appealed to families to send in an image of their lost loved one, so that they could be reproduced 'in presentation album form'.[2] Any remaining romanticism about the war had evaporated.

A tunneller on 'listening duty', featured on the bronze plaque, National Miners' Memorial, National Memorial Arboretum, Staffordshire. Brian Elliott

In a variety of service, many thousands of miner-casualties accrued from the major campaigns and battles: the Dardanelles and Gallipoli, Neuve Chapelle, Hill 60, Ypres and Loos. Recruitment from the mines stalled. The 'rush to join the Colours' waned despite continued boasts from the coalfield newspapers of the day. In the coal town of Barnsley, it took several months and renewed appeals in order to raise a second Pals battalion, quite a contrast to the few days it took to assemble the first. In Wales, a 'recruiting event' in the form of a boxing carnival in Cardiff Arms Park resulted in just thirty-eight volunteers from a crowd numbering 8,000.[3]

At home, Belgian refugees continued to arrive, accommodated in and around pit towns already reeling from food shortages, due to the German navy's attacks on merchant shipping. The bigger ships and Admiralty vessels depended on the supply of steam coal, produced by Welsh miners; and yet the workers and their families struggled to make ends meet, especially those

Private Francis Percy Warren

Getting compensation from colliery companies when a former employee was killed in action was far from easy. Percy Warren lost his life on 18 June 1915, aged twenty, only a few weeks after starting front-line service in the trenches. It took months and several appealing letters from family and friends for Mary, his widow, to receive any financial support. Official confirmation of his death was absent and then much delayed. Though fragmentary in places, the appealing letters to the army among his papers make sad reading.[4]

Originally from the Wigan area, after his family moved to south Yorkshire, at the age of seventeen, Francis joined the Territorials in Rotherham, in 1912. He worked at Rotherham Main and was a pony driver. After war was declared Francis entered the ranks of the 1st/5th York and Lancasters, sailing to France on 14 April 1915. Another interesting feature of his short life was that he wrote to his colliery manager – an action so unusual that it was published in the *Rotherham Advertiser*.[5] He informed him of the death in active service of a former workmate, and explains that many 'Rotherham Main men' were at the Front. Warren was interred at Y-Farm Military Cemetery, Bois Grenier, France (Grave G 24).[6]

Private Percy Warren. Brian Elliott

where one or more wage earner served overseas. The great gulf between the wealth of the mine magnates and their workers widened. Miners were paid on rates dating back to 1910, whereas the coal owners benefitted from a 50% increase in the price of coal since the start of the war.

'Their work in another guise': Durham miners at work at a tunnel entrance, one dragging a full crate of excavated earth and 'sapping', creating a covered trench, c.1915. Brian Elliott

A heroic colliery manager: Robert ('Bertie') Richmond Curwin Blair, DSO

Captain R.R.C. Blair, commanding 'A' Company of the 5th Border Regiment, and managing director of Whitehaven Collieries, was awarded the DSO for conspicuous bravery near Armentières, France. On the night of 27 September, after taking out a bombing party from his trench, Blair opened fire with his revolver when faced with a dozen of an enemy bombing party advancing a few yards away, 'dropping' four of them, including an officer. Three more Germans were hit by his fire when they began to escape, the action certain to have saved many British lives.

On another occasion, Captain Blair and one of his men crossed open ground under heavy fire in order to rescue two wounded Royal Engineers, an officer and corporal. The officer was dead but the corporal survived. Recommended for the VC for the deed, Blair, a qualified mining engineer, was a popular leader for the miner-soldiers under his command. Much respected, in 1910 he had been awarded the Albert Medal for his bravery in the Wellington pit disaster. Blair was presented with a 'sword of honour' by the Corporation of Whitehaven, in a ceremony where the widow of miner-VC winner Private Abraham Acton was also in attendance.[8] But, aged thirty-seven, he was fatally shot whilst in an action near Ypres, on 21 July 1916, and was interred in the Dranoutre Military Cemetery.[9]

Captain Robert 'Bertie' R.C. Blair, DSO. Colliery Guardian

One change became more apparent: existing miner recruits were being targeted for transfer or attachment to specific roles within the military. In February 1915, fifty of the first batch of Barnsley's first Pals battalion were seconded to the Royal Engineers, as part of one of the new tunneller companies. This arrangement was repeated elsewhere wherever former miners were in training for military service, or already serving overseas; and, as if to publicise this, Durham miner-recruits to the New Army were pictured in patriotic magazines 'trench-digging and sapping', doing 'their old work in a new guise'.[7] From October, the Derby Scheme (see p. 45) made miner-enlistment more difficult, though by no means impossible.

On the Home Front, the fall in coal production was a major pre-occupation of the government. Accordingly, in February, the Home Secretary, Reginald McKenna, established the Coal Mining Organisation Committee (CMOC), chaired by Sir Richard Redmayne, Chief Inspector of Mines. Its remit was 'to secure the necessary production of coal during the war'.[10] Closely allied to this was a committee set up two days later (feeding into the CMOC) on the retail price of coal. The CMOC proved to be a crucial body in monitoring and maintaining coal supplies, sensibly withdrawing an initial reference to suspending the 1908 Eight Hours Act (limiting underground work time for each shift) so as not to antagonise the unions. But it did flag up the 'cost to the nation' of miners' absenteeism and the importance of 'better productivity'. Try explaining that to most of the front-line miners of Britain who had to get coal by hand tools in low seams.

The main cause of decreased output was not 'shift-dodging' but the shortage of manpower due to the great loss of miners to the armed services. The CMOC found that after twelve months of the war, over 250,000 miners had left the industry but only 75,000 replacements were found, a 16% net loss of labour. During the same period the output of coal had fallen by almost 11%.

Among the remedial measures recommended by the CMOC was an increase in employment of women at collieries. Despite much union opposition, women's work was deemed by the committee as 'not degrading', indeed was felt to be 'healthier' than in factories. Employment of women at collieries actually increased by a significant number: from 6,554 in 1914 to 8,312 (27%) in 1918, mainly by recruitment in Lancashire and Scotland.[11] Any suggestion, however, to employ more boys underground was ruled out due to anticipated union opposition.

Keep the Home Fires Burning,
While your hearts are yearning,
Though your lads are far away
They dream of Home.
(From the revised (1915) edition of Ivor Novello and Lena Ford's popular song).

News about the recruitment and the 'war-skills' of 'military miners' continued to be promoted in the coalfield presses, even in south Wales, despite its radical and nonconformist traditions. Len Llewellyn, representing the large Cambrian Combine coal ownership, said that the Welsh collier showed 'as much pluck on the battlefield . . . [and] especially at times of explosions, [so] they would give a good account of themselves [in the war]'.[12] But 'patriotic' results could not be argued against. A proposed battalion made up of rescue men, by a Mr G. Blyth, of Rugby, included applications from six Welsh mine rescue teams.[13] The moderate Rhondda miners' leader David ('Dai') Watts Morgan, who had joined the 10th (1st Rhondda) Battalion of the Welsh Regiment at the outbreak of war, was one of the louder voices for recruitment, bemoaning any objectors as 'peace cranks'. Soon promoted to Captain (and then Major), he began the process of raising a Pioneer battalion;[14] and sure enough, by March 1915 had sworn in 4,510 men, mostly miners, to various regiments.[15] The Welsh industrialist and Liberal MP David A. Thomas, owner of the giant Cambrian' group of mines, a survivor of the sinking of the 'peace ship' RMS *Lusitania* on 7 May, was a key voice regarding domestic and international coal supplies. Sent on a munitions mission to the USA by Lloyd George, Thomas' efforts were later (1918)

The Group ('Lord Derby') Scheme

Introduced by the government in October 1915, men between the ages of 18 and 41 were *asked to attest their willingness to serve*. Miners were classed as 'starred' workers, excluded from service because of the vital importance of their jobs for the nation. A 'Derby notice' posted at every pithead included instruction that workmen offering themselves as recruits would only be accepted on the condition that they return to their collieries 'until called upon'; and would be 'given armlets' in recognition that they 'are willing to obey the call'.[16] The order applied to all miners working underground as well as specialist surface workers, identified as 'winding enginemen, pumpmen, weighmen, electricians, fitters and mechanics'.

The coalfields were divided into twenty-three recruiting districts, each with a tribunal to advise the recruiting officer on individual cases regarding acceptance or exemption. Yet many miners, not just clerks and miscellaneous pit-top workers, fell through the Derby net and enlisted.

rewarded with the title of Viscount Rhondda. About the same time, Dai Watts Morgan got a DSO.

In the huge Yorkshire, Nottinghamshire and Derbyshire coalfield more accurate figures for the enlistment of miners began to appear in publications. The Nottinghamshire Miners' Association confirmed that 5,400 of its 33,000 members (16.3%) 'have joined the Colours'.[17] Recruiting varied in Yorkshire, which averaged 11.5%, according to its mines inspector, Thomas Mottram. Yorkshire miners were targeted by the Royal Engineers, 'for sappers and tunnellers', volunteers said to be coming forward on a daily basis.[18] At Chesterfield Town Hall, sixty men specially selected for tunnelling on the Western Front attested from Derbyshire collieries.[19] In Doncaster, with an eye on their tunnelling potential, young unmarried miners were 'being canvassed for enlistment' irrespective of their 'starred' status.

In the North East, William Straker, secretary of the Northumberland Miners' Association, reported that 14,000 miners had enlisted from his area, including about 5,000 non-members, representing about 25% of the total workforce, a huge response. Straker was particularly praiseworthy of the Ashington district, and singled out North Seaton, where 668 out of a total membership of 882 (a massive 76%) 'have joined the Army or Navy'. Later, Straker admitted to a problem: finding sufficient labour 'to work the pits' was 'becoming increasingly difficult'. He also made the point that the remaining coalface men were either over age or unfit for military service; and that the underground haulage of the coal was now in the hands of 'older men or mere youngsters'.[20] In the Durham coalfield, Ryhope Colliery was reported to have had an exceptional record of recruitment, 58% of its miners of military age having enlisted.[21]

Durham miners were often mentioned in the press in 1915 because of their outstanding acts of bravery whilst on military service. Several Thornley Colliery men had been specially selected for 'mining operations' in France. One, Private John Henry, a 'bomb thrower' in the Irish Guards, was awarded the DCM 'for killing a bunch of Germans' on the night of 2-3 August.[22] Henry's action took place at the 'mining hotbed' of Givenchy when he was threw grenades from a forward sap at the edge of an exploded mine, getting wounded in the process. The small French village was, according to military historian Phil Tomaselli 'a target for crushing bombardments, infantry assaults and subterranean warfare' in 1915-16.[23]

Outlining the many advantages that miners had in regard to army service, a reporter for the *Durham Chronicle* referred to their 'uncanny ability to see clearly in what to ordinary persons is almost absolute darkness', the men ideal for 'the conduct of night attacks, patrols, and reconnaissances'.[24]

James Morrison, 8th Durham Light Infantry, DCM

No sooner had the *Durham Chronicle* celebrated the award of a DCM to former Hetton miner James Morrison than it reported on the same man's death.[25] The award citation read:

> For Gallant conduct and resource near Ypres on 25 April 1915, when the parapet of the trench had been blown in and all the Officers and Non-Commissioned Officers in the vicinity had been killed or wounded, in organising the men near him with great coolness and continuing to fight until ordered to withdraw.

Sergeant James Morrison, DCM. Durham Chronicle

In a letter home, Morrison described the event in his own, rather matter-of-fact words: 'I was the last man left in the trenches with Captain Bradford. I killed thirty Germans and a German officer, and hope to be back soon, and do better. I hope to return for the Durham Big Meeting.'

James's actions were extraordinary, in the tail end of the (Second) Battle of Ypres. The Big Meeting that he referred was the annual miners' gala, held in July. An Acting Sergeant, Morrison was conveyed to 3 Casualty Clearing Station, suffering from severe shrapnel wounds to his head and face, according to a letter sent to his wife, by the Church of England Chaplain, A.F.G. Fletcher. Aged twenty-seven when he died, on 26 June 1915, Morrison was interred in Bailleul Communal Cemetery Extension; and is remembered on West Rainton's war memorial, in County Durham.

Recruitment of Scottish miners to the military continued apace, local

Private Thomas Kenny VC, Durham Light Infantry[26]

Thomas was among the 13th DLI landing at Boulogne on 26 August 1915, bound for the trenches of Flanders, and many of his soldier pals were former miners. At the Front he served as an observer to an officer, Lieutenant Philip Brown. An Oxford graduate, Brown preferred working as a tutor for the Workers' Educational Association in Durham pit villages than earning a lot more as a lecturer elsewhere, so was much respected by the miner-solders. In a letter home, he described how 'his man', on the very first night of the BLI's trench duty, had carried on his back a wounded soldier from a shallow trench.

Company Sergeant Major Thomas Kenny wearing his VC. Brian Elliott

A few weeks later, on 4 November 1915, Kenny was the subject of an amazing act of gallantry, in an action at La Houssoie, that won him the Victoria Cross. Shot through both thighs, Lieutenant Brown was carried under heavy fire by Kenny through No Man's Land into an old trench. Leaving Brown as comfortable as possible, the now battered and exhausted Kenny managed to trudge and crawl to the battalion listening post, where Captain White was informed of the grim situation. Despite his own dire condition Kenny led a small rescue party back to where he had left his officer, the stretcher-bearers reaching him despite heavy rifle fire and exploding grenades. Sadly, Brown died from his wounds on the way to a dressing station.

As a young man Thomas Kenny worked as a quarry labourer in South Wingate, County Durham but by his late twenties, earned more money as a miner, at Wheatley Hill Colliery. He certainly needed a decent wage in order to support his wife, Isabel, and six young children, crammed into a terraced house in Walker Buildings; but he decided, aged thirty-two, to join B company, of the 13th DLI. Private Kenny, to paraphrase his own later words wanted 'to do my duty'.[27]

Thomas Kenny returned to front-line action after his VC investiture at Buckingham Palace, rescuing a Sergeant on the Somme in 1916, a deed in which he received a serious gunshot wound to the leg. At home, among several presentations to him was 'a purse of money' subscribed for by his Wingate Colliery workmates.[27] Promoted to Company Sergeant Major, the end of the war was a sad time for the family, his brother Hugh killed in action in August 1918.

After discharge, Kenny returned home and resumed life as a miner, working at Wingate and Wheatley collieries, latterly as surface worker after an accident in 1944. Barely enjoying retirement, he passed away on 24 November 1948, and was interred in an unmarked plot in Wheatley Hill Cemetery. Kenny had the distinction of being the first (of six) DLI men to be awarded the VC during the First World War. Half of this distinguished group (Kenny, 1915; Heaviside, 1917 and Young, 1918) were former coalminers.[28]

newspapers praising colliery villages for their patriotism. Steeland (Fife), only had a small population of 300 but 60% of 'eligible men' had become 'soldier miners'.[29] A growing roll of honour compiled in the Lothians, of men from the miners' union serving in the army, had 2,000 inscriptions.[30] Three Lanarkshire miners' leaders visited the Front in France and Flanders in August, the welcoming officers informing them of the great importance of Scottish miners as 'trench-diggers'.[31]

In the Midland and Southern coalfields, the sparsity of recruits from Cannock Chase was put down to many of the miners living in rural locations. Accordingly, a 'Pals-style' battalion was proposed, so that the men could enlist together rather than be 'among strangers'.[32]

In the new Kent coalfield the average recruitment rate per pit was high, 28%. Worried over production and profits, the colliery companies pleaded that no more men should come forward as it would harm winter coal supplies.[33]

Coal production

Fearing coal shortages, a somewhat bizarre suggestion was made in committee that men 'unfit to enlist' could be sent to the mines 'to make up the numbers'. The more sensible delegates poured cold water over this inept proposal, presumably because they were more aware of what mining coal really involved. Another silly idea, was to transfer married men from other parts of the country – at the Government's expense – to south Wales, but again common sense prevailed.[34]

The *Colliery Guardian* suggested that miners were 'not producing as much as they could', a contentious view hotly disputed by the miners' unions.[35] Maintaining coal supplies had become such a big issue that when the MFGB convened a meeting in March, their agenda including pleas from Kitchener and Redmayne, asking the membership to shorten their Easter holidays from three to two days.[36] Generally speaking, the request was adhered to, indeed in parts of Yorkshire some of the men took just one day's holiday. Astonishingly, a million tons of coal was 'saved' through this 'sacrifice'.

The introduction of female labour for pit top work was mooted, the strategy being that boy miners could then be released to work underground. This scheme came to an abrupt end in County Durham, at Littleburn Colliery, the employment of twenty-four females recruited to work 'on the belts' (the screens), was postponed following a strong protest by the DMA.[37] At a meeting near Maryport, though out of his union area, William Brace, president of the Welsh Miners' Federation, urged the miners to 'put every ounce of energy into their work' as Cumberland coal was important in the making of high explosives.[38] Brace was well aware of the bargaining power and strategic importance of Welsh 'steam' or 'Admiralty' coal.[39] In Yorkshire, some mine managers dealt with their labour shortages by 'knocking off' one of the three shifts; and at some pits underground development work was also put on hold.[40]

Miners taking unofficial days off work was a problem that inflamed union-management-owner relations throughout 1915. Coal companies began to prosecute some of their workers. At Grassmoor, Derbyshire, twenty-four men were off work for an accumulated total 215 days in a single month. The Grassmoor Colliery Company stated in court that 493 of their workforce had 'patriotically enlisted'. The outcome was that the twenty-four absentees were fined, bar a few with 'genuine illnesses'.[41]

Belgian refugees

A stance was taken by the Home Office against any Belgian refugees working in the mines, unless there was an interpreter with them, or they had a reasonable command of English. Those refugees said to have found employment in Doncaster area pits were, therefore, in danger of being withdrawn completely or redeployed on the pit-top. Redmayne, the chief mines inspector, reporting to the new committee on the employment of Belgium refugees, said that it was 'highly dangerous' to employ underground workers in mines who were unable to understand instructions and rules.[42] Later, acting on a decision by the Yorkshire Miners' Association, the employment of ten refugees who had been working at Barnsley Main Colliery was terminated by the owners, and transferred to their 'headquarters' in London.[43]

Belgian refugees fleeing the invading German army. Taylor Library

The Goldthorpe riots

In the wake of the sinking of the *Lusitania*, many public order disturbances erupted throughout Britain. An extreme example took place in south Yorkshire, in the large pit village of Goldthorpe, midway between Barnsley and Doncaster. Here, on the night of 11 May, Schonhut's pork butcher's shop was subjected to protests by a large and hostile crowd. The disorder was so serious that the police were unable to quash looting and attempts to set fire to the premises. The day after, another shopkeeper, John Bakewell, rumoured to be related to the Schonhuts, decided to barricade the windows of his extensive premises after evening closing time, about 9pm. Like a red rag to a bull, his window protectors were then attacked by 'an unruly mob'. Stones were thrown, glass broken and lights put out. Faced with looting and an invasion of his house, Bakewell resorted to use a firearm, shooting 'over the heads' of the incomers. One man, John Eades, was so gravely wounded that he died several days later but, to the last, refused to blame Bakewell for his demise.[44] Seventy-two people were arrested and twenty-one of them charged with 'riotous assembly', a very serious accusation. Most were miners, probably from the local pit, Hickleton Main. In the West Riding Crown Court at Leeds, a three-day trial resulted in seventeen of the men given custodial sentences ranging from two to fifteen months.

Goldthorpe, which had already provided over a hundred Kitchener volunteers, was said to 'close in' on itself afterwards, full of resentment due to alleged police brutality during the riots and a high-profile police presence afterwards.[45]

War ambulances

Miners combined with pit owners in the provision of ambulances for the Front, one of the most underrated contributions of the coal industry and the war. The actual funding process was led by a remarkably energetic 36-year-old Nottinghamshire coal owner, Dennis Bayley (Henry Dennis Readett-Bayley). Bayley's ambitious target was to raise sufficient cash for the Red Cross and St John's Ambulance service to be given two conveys, each consisting of fifty fully equipped ambulances, supported with touring cars and a repair vehicle. By June, the 'Bayley initiative' had gained support of the Derbyshire and Nottinghamshire miners' associations,[46] and it was not long before it became a coalfield-wide movement, attracting the support of professional bodies such as the Mining Institute (NEIMME).[47]

By autumn 1915, the Joint War Committee of the St John Ambulance Society and the Red Cross were able to take delivery of the new vehicles. On 14 October, the King and Queen inspected a section of the Derbyshire and Nottinghamshire coal owners' and coal miners' ambulances, many thousands of miners having paid 6d (2.5p) a week from their wages. A similar-sized convoy was inspected by the King at Buckingham Palace

Ambulance fund-raiser Lieutenant-Colonel (RAMC) Henry Dennis Readett-Bayley. Brian Elliott

'War ambulances' paraded at Brierley Rescue Station, near Grimethorpe and Monckton collieries, Barnsley. Brian Elliott

in December, courtesy of the Lancashire and Cheshire coal owners and miners.

The Dennis Bayley Fund raised an astonishing £700,000 (c.£42m today) by 1918, enabling the commissioning and provision of hundreds of ambulances, hospital boats and hospital beds in the pit towns and districts of Britain.[48]

A Lieutenant-Colonel in the RAMC, Bayley was mentioned in dispatches, and as a 'fourteener' got the Mons Star. In fraternal appreciation, France also awarded him the Croix de Guerre and the Legion of Honour; and back home he was created a KBE (Knight Commander of the British Empire).

Bayley praised the 'ambulance men' who staffed his new vehicles, most of them former miners serving in the RAMC and ASC. Addressing the Durham Miners' Association in August 1915, this is what he said:

> The colliers have played a most important part in the war ... Ninety per cent of St John Ambulance men are ... from the collieries – a tribute to coalminers which will go down in history, showing that the spirit that enabled them to prepare and assist their injured brothers in times of peace helped them in a similar manner in times of war [applause].

Mines inspectors' concerns about the war and coal production

The great exodus of miners from the collieries to 'join the Colours' during 1914-15 was a feature of the mines inspectors' reports for 1915. Overall, the number of employees 'in and about the mines' had fallen to 953,642, a reduction of almost 16% (1 in 6 jobs); but some collieries fared far worse than others, losing as much as half of their workforce.

Hugh Johnson, the Midland and Southern inspector, expressed concern that the loss of labour to the military necessitated the drafting into the mines of 'a considerable number of men from other industries' and 'a number of young persons under sixteen years of age', hardly equal to the skill and experience of regular miners.

All divisional inspectors acknowledged the 'effect of the war' on reduced manpower and production. Safety, however, despite the annual number of accidents actually increasing (to 1,297, from 1,219 in 1914) was not seen to have been affected by falling numbers. However, serious accidents at Podmore Hall, Hem Heath and Exhall collieries in Staffordshire (nine, twelve and fourteen fatalities respectively); and at Bentinck in Nottinghamshire (ten deaths) were hardly unimportant occurrences.[49]

The mines inspectorate itself suffered from a serious reduction of staff due to the war. Twenty inspectors had enlisted, two of them getting killed in action (Robinson and Foot) and two others (Clive and Spikin) badly wounded. All but one of the remaining enlistees remained in service, three of them getting the Military Cross.[50]

Miner-VCs

For acts of extreme bravery during 1915, ten former miners were awarded the Victoria Cross, including Durham Light Infantry man **Thomas Kenny**, already referred to above.

Lance Corporal **Wilfred Dolby Fuller**, a 21-year-old Grenadier Guard, was the first, achieved at Neuve Chapelle on 12 March. Single-handedly, he chased and got a cohort of almost fifty Germans to surrender, after killing their leader with a grenade when they were attempting to escape along a communication trench. Son of a Mansfield colliery deputy, Fuller was a pit pony driver when he enlisted in 1911, and was in France from 8 November 1914, with the 1st Battalion of the Grenadiers.

He was given a public reception in honour of his gallantry, in his home town of Mansfield in April 1915, the Mayor alongside several dignitaries including the principal sponsor, mine owner Sir Arthur Markham, presenting him with an Illuminated Address and a gold watch. Fuller was also heartened by the presence of the Mansfield Colliery band and a procession of miners preceding the civic carriage on the way to the ceremony.[51] Later, in August 1915, he was awarded a prestigious Russian medal, the Cross of the Order of St George. Discharged as 'physical unfit' from the army at the end of October 1916, he was unable to resume pit work on a regular basis, but found employment with the Somerset police.

Artist's impression of Wilfred Dolby Fuller's action at Neuve Chapelle.

British troops occupy the captured Turkish trenches in Gully Ravine, Gallipoli. Taylor Library

William Stephen Kenealy.

Two Lancashire miners got VCs for one of the most illustrious actions of the Gallipoli campaign, during the landings at W. Beach, Cape Helles, on 25 April 1915. The older of the two, 29-year-old **William Stephen Kenealy**, was Irish-born, his father finding work at Bryn Hall Colliery after service with the Royal Irish Guards. William was a boy miner prior to a seven-year stint with the Lancashire Fusiliers, enlisting in 1909 and re-joining his regiment at the start of the war. Kenealey showed outstanding bravery as a 'runner', delivering messages under heavy machine gun fire and crawling through entanglements of wire in the process, enabling the beach to be won. Afterwards promoted, he survived the three battles of Krithia which almost wiped out his unit but was mortally wounded in the Battle of Gulley Ravine on 29 June 1915. The news of his death took four months to reach his family who had anticipated a celebratory homecoming for their 'VC hero'.

John Elisha Grimshaw worked alongside his father as a carpenter on the pit top near his home at Abram, Wigan. It was during this employment that in 1910 the worst ever death toll in a Lancashire colliery occurred, not far away at the Pretoria (Hulton) Pit, 344 men and boys killed. Two years later, aged nineteen, John escaped from the potential dangers of mining to join what he may have thought as the relative safety of the Lancashire Fusiliers. Instead, on 25 April 1915, as a signaller in C Company, he had to crawl and cut through barbed wire under a relentless hail of bullets on a foreign beach, the most hellish of experiences imaginable.

Kenealy and Grimshaw were two of a small number of soldiers chosen by the Fusiliers for the VC, the group soon to be known as 'the six VCs before breakfast'. Grimshaw survived Gallipoli, though not without mishap, buried in a shell blast and suffering from frostbite. His Victoria Cross was, belatedly, gazetted and presented in 1917, by which time he was a Sergeant.

John Elisha Grimshaw pictured after collecting his VC in 1917.
Daily Mail

William 'Willie' Angus was brought up in the small Lanarkshire town of Carluke and followed his father's footsteps, finding work at a local colliery. As a young man he was so good at football when playing for the town team that he was taken on as a professional by Glasgow Celtic, though over two seasons (1912-14) he never made the first team. A Territorial, he joined the Army at the start of the War, the Highland Light Infantry's 8th Battalion, enlisting on 9 September according to the remnants of his military records. Willie embarked to France in February 1915, and was soon in front-line action at Neuve Chappelle, suffering gunshot wounds to his right leg on 18 May. Subsequently hospitalised at Wimereux and Rouen, he was still able to rejoin his unit by the month end, returning to the front line with the 8th Royal Scots at Givenchy on 10 June, where his VC action took place two days later.

There are numerous accounts of Angus's heroic deed, Batchelor and Matson and Tomaselli providing particularly well researched descriptions.[52] Angus suffered wounds to his head, body, limbs and left eye which had to be

Studio photograph of 'Willie' Angus as a Celtic FC player.
Brian Elliott

removed, and also his right foot. The Highland Light Infantry's commanding officer, Colonel Gemmill, in a letter to Angus's family, described his 'heroics' in glowing terms, including an extreme comment that 'No braver deed was ever done in the history of the British Army'. Immediately recommended for the Victoria Cross, gazetting took place on 2 July 1915, the citation as follows:

> For most conspicuous gallantry and devotion to duty at Givenchy, on 12 June 1915, in voluntarily leaving his trench under very heavy fire and rescuing an officer who was lying within a few yards of the enemy position. Lance Corporal Angus had no chance of escaping the enemy's fire when undertaking this very gallant action, and in effecting the rescue sustained about forty wounds from bombs, some of them very serious.

In the wake of the VC presentation at Buckingham Palace, Angus was given a hero's welcome in his home town, as well as at Celtic FC. Among his many testimonials and gifts was one that he appreciated more than any of the others – a gold watch from James Martin, the officer who he rescued. Discharged from service after hospitalisations, William was not fit enough to return to mining.

One of the first Scottish miners (and indeed the first British professional footballer) to be awarded the VC, Angus remained fiercely proud of his home town, Carluke, where he died in 1959.

The other miner-VCs are James Upton (Sherwood Foresters), Frederick Barter (Royal Welsh Fusiliers), William Mariner (King's Own Rifle Corps), Robert Dunsire (Royal Scots) and Oliver Brooks (Coldstream Guards).

The first tunnellers: Lance-Corporal J.H. Davies, 171 Tunnelling Company RE

The Tunnelling Company of the Royal Engineers (171TC) spearheaded military mining in 1915 when supporting the infantry on Hill 60, Ypres, in April 1915. 'Clay-kickers' (former sewage workers), worked alongside miners, mixed contingents from several coalfields, but most notably from south Wales.

There were several early casualties. Sapper John ('Johnnie') Henry Davies experienced a horrible death on 20 June 1915. A blast from an underground explosion ripped through the excavated gallery where he was listening to Germans working nearby, collapsing the workings and 'burying him alive'. His demise was far removed from the 'instantaneous' and 'looked as though he was asleep as he lay on a stretcher' account of his death subsequently published in the *Barnsley Chronicle*.[53]

Davies was one of fifty pitmen seconded from the 13th Service Barnsley Battalion (Barnsley Pals) into the Royal Engineers. When he enlisted in September 1914, he was working as a miner at Wharncliffe Woodmoor Colliery. Married, with three small sons, when he was dispatched to Belgium, Johnnie Davies and a few of his Barnsley Pals were in the company of miner-enlistees from other coalfield areas, a mix of Geordie, Welsh and Black

Sapper John Henry Davies.
Barnsley Chronicle

A panoramic view of Hill 60, Ypres.

Country men. All had minimal military training but had lots of experience of underground work, perfect (at least in theory) for digging under Hill 60. The latter (or perhaps more aptly 'Murder Hill') was a man-made feature, composed of dumped spoil from the making of the Ypres-Commines rail link and canal cutting. Davies and his collier mates faced mining in the worst conditions imaginable, improvising with short lengths of timber to support the walls and roofs in tunnels less than a yard square. Working in pairs, the tasks of the tunnellers are well described by author Jon Cooksey:

> The two of them would huddle together in the red, oxygen-starved glow of a candle. One would laboriously pick and pry into the heavy oozing clay with a point of a bayonet or bare hands. Another would scoop up the dripping waste and place it into sandbags or, in some cases, onto small trolleys. It would then be pulled towards the mine entrance and distributed carefully on the surface to prevent detection by German observers.
>
> During the eight-hour shifts the men would spend long periods working in painful cramp inducing postures, soaked to the skin in a vile smelling concoction of their own sweat suffused with muddy water. Whereas once they had been covered in coal dust from the mines of Barnsley, they were now plastered from head to foot in a layer of slimy sludge.[54]

At 7pm on 17 April all the hard work of the miner-tunnellers reached a climax when a massive blast, from six chambers tightly jammed with explosives, was ignited under the German infantry, destroying a labyrinth of workings under the hill. Any strategic outcome was short-lived, as Hill 60 was ascended yet again by the Germans within a few weeks. Davies and 171TC's further underground explorations were seriously affected by bombardments of thick billowing gas seeping into the workings. Many hundreds of 'gassed casualties' accrued each week, including Davies who was admitted to a casualty clearing station, prior to recovering at base.

Parts of Davies' last letter home, dated 17 June 1915, three days before his death, summed up thoughts felt by so many of the tunnellers on Hill 60 and elsewhere:

> The whole place looks grand in summer weather, but the boom of the guns and the constant traffic of men and horses spoils the peacefulness of Belgium. I always tried to lead a good life at home, but I think I shall lead a better one after this, as the war has taught me and a good many more, a lot we shall never forget. If anything was to happen to me I should not wish you to fret at my loss but rather thank God that I had done my little bit for love and freedom, King and country.[55]

J.H. Davies is commemorated on Panel 9 of the Menin Gate Memorial to the Missing, Ypres.

Tunnelling Companies of the Royal Engineers

Military mining, offensive and defensive, was the most extreme form of work imaginable, far more perilous than extracting coal under regulated conditions. The main cause/fear of death and injury included burial by cave-ins, drowning by sudden inrushes of water, asphyxiation due to lack of air when at work in very cramped galleries – and, increasingly, gas poisoning. The blowing of an enemy mine in the direction of a working or listening area brought terror, burial and almost certain death. After 'break-ins' and 'break throughs' it was by no means exceptional for hand-to-hand fighting to take place, such was the proximity of operations.

John Griffiths (his surname changed to 'Norton-Griffiths' in 1917), a Conservative MP and a former captain in the South African War, had considerable experience as a public works contractor. One biographer summed up his abilities in terms of him 'having the strength of a prize-fighter and the temperament of a guerrilla leader'.[56] Under the badge of the Royal Engineers, Griffiths adopted methods he had used when directing his sewer workmen, famously known as 'clay-kickers', and referred by himself as his military 'moles'. His proposal to establish specialist tunnelling units gained the approval of the War Office, thanks to his sheer enthusiasm, persistence and charisma.

Major John (Norton-) Griffiths.
Wikimedia Commons

The first seven of Griffiths's tunnelling companies, allocated numbers 170-177, were ready for deployment by mid-February 1915, and most of the others (178-185 and 250-258) formed by the end of the year. By the time the last British TC was assembled, the 256th, in July 1916, there were about 25,000 specially trained British tunnellers, mostly former miners. Numbers were almost doubled with the addition of 'attached infantry', an essential labour force responsible for a myriad of support jobs, from removing spoil and pumping water to fetching and carrying mining equipment.

Where opposing trench lines were within 'tunnel reach', military mining was a 24-hour, 365-day operation. Soon, parts of the Western Front consisted of complex warrens of underground galleries below crater-strewn stretches of No Man's Land, accessed by shafts of varying depths and circumferences.

Shaft-sinking was a notable skill of specialist British miners, who were able to combat sandy and wet geological conditions by the use of steel linings ('tubbing') for the sides of otherwise unstable and dangerous descents.

Enoch Dalton, a most reluctant hero[57]

For some miner-volunteers, joining the Colours was not an easy decision but at least they had a choice. A significant number of others, the Reservists, had little or no option to respond, irrespective of personal circumstances. Enoch Dalton was thirty-three years old, happily married to London-born Ann when war was declared, and they had four small children. The Daltons had moved to Maltby, near Rotherham, in the latter part of 1911, where Enoch, an experienced miner, was able to find work at a large new mine, Maltby Main. Enoch's job was a good one as he worked as a hewer. He may have been one of a cohort of pioneer miners, set on to tunnel towards the main Barnsley seam, extracting any loose coal which was then hauled to the surface. It was during this task that three men lost their lives.[58]

Sapper Enoch Dalton DCM (and Bar). Jane West

A six-footer, strong, fit and well, Enoch was well suited to life as a Guardsman since as a young man, in the summer of 1899, he had escaped from a previous pit and joined the 3rd Battalion of the Coldstreams for a three-year stint of duty, experiencing barrack life in London for the first time. It was on sentry duty that the 21-year-old Private caught the eye of a young spinster, Ann, the couple marrying at St George's, Hanover Square, on 16 December 1901. Life as a coalminer resumed, principally at St John's Colliery, Normanton in west Yorkshire, but Enoch remained a Reservist.

On his thirtieth birthday just prior to his move to Maltby, Enoch Dalton took the fateful decision of re-engaging in the Army Reserve. The outcome was, quite suddenly, having to leave his young family to retrain for combat with the Coldstream Guards – only two days after war was declared. In a letter sent to a friend on 17 August 1914, Enoch expressed views more akin to that of a pacifist, the reflections of a thoughtful and sensitive individual, far removed from the stereotypical view of a burly coalminer:

> Today I am merely a machine of destruction, ready at the word of command to slay those whom I would prefer to shake hands with ... and I hope this war means the Death to Autocracy and military despotism which seems only to flourish on the Blood & Sinews of the masses ...

By September and through to Christmas and the spring of 1915 Dalton was on and near the front line in France and Flanders. Here, he experienced some of the most horrible conditions, sights and sounds imaginable. Enoch's abbreviated writings leap from the pages during the (Second) Battle of Ypres and its grim aftermath: 'shell fire', bullets through buildings, digging trenches which were 'frosty' at night and 'rotten' and 'untenable' by day' among his comments. The sight of dead comrades and a 'Hell Day' (23 October) of horrific shelling completed his baptism to active service. In the relative safety of the reserve trenches a couple of days later, Enoch noted that an exploding shell caused twenty-five casualties, and his commanding colonel, and another officer, Captain Monk, were killed.

Combat and conditions worsened even more after Enoch and his comrades reached Givenchy on the La Basse Front on 1 February 1915. On arrival he saw thirty-four 'dead Highlanders' and 'piles of dead' soldiers, counting fourteen in one heap and seventeen in another. Based at Bethune, next day his company faced intense German fire power when they returned to their trenches, shrapnel raining down on Enoch, one shell blowing his helmet off and an explosion scorching his legs and causing temporary blindness. On 6 February, another of Enoch's friends was killed and several other comrades wounded during a bayonet attack on the Germans.

There was little left. Two more of the company's soldiers were killed, a comrade had his finger shot off and his best friend William Greer suffered a gunshot wound to his back. Excruciatingly hard night marches must have reinforced Enoch's thoughts and feeling about the futility of war and indeed combat in general. When not digging trenches he endured 'frosty' and stench-laden conditions, described as 'untenable' during onslaughts of German shell fire.

By March, Dalton and a few other specially-chosen former miners had merged with the Griffiths volunteers as part of a new unit, the 170 Tunnelling Company (attached to 11th Field Company), the first of eight specialised units of the Royal Engineers to be set up for deployment on the Western Front between La Basse in northern France to Ypres in Belgium. As an experienced hewer, Enoch was a perfect for the tunnellers and may well have volunteered for his new role, under the 170TC's commander, Captain Preedy.

Six months into active service, Dalton was an integral part of the subterranean war that turned the open land between opposing trenches and along about a thousand yards of No Man's Land into a mass of craters. His diary entries become less detailed, perhaps reflecting the covert operations of tunnelling. Preedy's 170TC, liaising with Norton Griffiths, were involved in a fifteen-day operation from a sector known as 'Brickstacks', driving a pair of tunnels – in very difficult ground conditions – to reach within a few feet of the German lines.[59] Dalton was at the forefront of the operation, working in galleries barely a yard high and about six inches less in width. Charged with gunpowder and guncotton, mines were blown in the tunnels at daybreak on 3 April. On this day Dalton's diary records '80yds of enemys [sic] trench completely buried' and 'others [retreating Germans] shot down as they ran away'.

Enoch may have empathised with the comments of the poet Robert Graves, a young officer serving with the 3rd Battalion Royal Welch Fusiliers, who in his autobiography *Goodbye to All That* referred to the 'constant mining' taking place in the Cambrin-Cuinchy sector during 1915.[60] Graves explored first-hand some of the claustrophobic, dark and stuffy subterranean passageways in which the tunnellers had to create and squat, listening in silence like trap-door boys in a Victorian mine:

> We had the prospect of being blown up at any moment . . . A duel of mining and counter-mining had been going on . . . At the end of the gallery I found a Welsh miner on listening duty, one of our own battalion, who had transferred to the Royal Engineers. He cautioned us to silence. I could distinctly hear the Germans working somewhere below us. He whispers: 'So long as they work, I don't mind. It's when they bloody stop!'. He did his two-hour spell by candle light in the cramped and stuffy dead-end, reading a book.[61]

On the morning of 14 April at Cuinchy, many of Enoch's colleagues were wounded after they had exploded a mine, and for the first time he begins to record 'listening duty' in his diary, part of the cat-and-mouse activities now common on both sides. An extraordinary act of bravery followed, on 21 April, resulting in Dalton being awarded the DCM. Gazetted on 30 June 1915, the citation summarised his action as follows:

> For conspicuous gallantry and good work under very dangerous conditions in charging and tamping mines within five feet of the enemy's mine. On the 21 April, 1915, at Cuinchy, he crawled forward and surprised three Germans behind a sandbag barricade, driving them out and thus enabling a charge to be fired and the gallery destroyed.

A major breakthrough for tunnellers such as Enoch working in the Ypres Salient occurred on 6 May when at Cuinchy Lieutenant Leeming and his section of the 170TC succeeded in sinking a timber-lined shaft through two yards of loose sand to reach the underlying clay. This action was so significant that it 'guaranteed that comprehensive deep British tunnel systems could be secretly established . . .', Norton Griffiths insisting that every TC commander should 'get into the clay'; and within twelve months there was 'almost complete British underground dominance at Ypres'.[62]

On 9 June, Enoch sent a postcard to his wife and family informing them of his DCM award. Only twelve days later, he was involved in a further act of bravery that got him a second DCM (a Bar), an extraordinary achievement. Gazetted on 5 August 1915, the citation outlines what appears to be a double act of bravery, and one that might well have been considered for a VC:

> For conspicuous gallantry and devotion to duty at Cuinchy, on the night of 21 June, 1915, in going down a mine, and assisting in the rescue of four men under circumstances of great risk. On the morning of the 22nd June the enemy exploded a mine, entombing about nine of our men, and this Non-Commissioned Officer went down the mine time

after time, with the greatest courage, and assisted in bringing out all the men. He was badly affected by the poison fumes.

Despite a need to 'rest up' and get some hard earned leave, Dalton's unselfish actions continued when he rescued several trapped Russian soldiers. For this he was awarded with their prestigious St George Medal (4th class), a rare honour for a British soldier.

After months of winter work in sodden trenches, hours of muddy marches and work in the cramped mines of Givenchy and Cuinchy, Enoch developed rheumatism that was so painfully debilitating that in the summer he was sent back to 'Blighty' for treatment and rest in Stoke War Hospital.[63]

Enoch Dalton's awards while serving as a Royal Engineers tunneller were acknowledged locally in November 1915, at a public meeting in Maltby 'picture house' when he was presented with an Illuminated Address and 'a wallet of

Edmund Blunden at Cuinchy

One of the finest poets of the First World War, Edmund Blunden (1896-1974) was a young officer in the Royal Sussex Regiment, serving at Cuinchy at the same time as Enoch Dalton. He was badly affected by the brutal destruction of the countryside which had become nothing more than a 'slaughter yard', referring to the 'brick-stacks' as 'historic strange monuments'. Blunden's retrospective writings about Cuinchy, include this comment, at an event in April 1916 which resulted in sixty casualties:

Edmund Blunden. Public Domain

> It was as dirty, bloody thirsty and wearisome place as could be found in ordinary warfare; many mines had been exploded there, and tunnelling was still going on. We had scarcely found out the names of the enemy trenches . . . when midnight was suddenly maddened with the thump and roar of a new mine blown under our front companies. The shock was like a blow to the heart; our dugout swayed, there were startled eyes and voices.[65]

On 7 August, a few days prior to Enoch Dalton's death, Blunden was in Givenchy, tasked with overseeing a major repair of a sap close to the Red Dragon Crater (named after the shape of the Royal Welsh regimental badge) where on 21-22 June a massive underground mine was exploded by the Germans resulting in the death of five soldiers from the 254TC, including Sapper William Hackett who got a posthumous VC for one of the most courageous rescue attempts in the Great War.[66] Not far away, Enoch would have been well aware of the huge explosion and earthquake, described by Robert Graves, as 'the biggest mine blown on the Western Front so far'.

notes' (cash) by the president of the Maltby branch of the Yorkshire Miners' Association. In his thank you speech Enoch appealed for mineworkers 'to put aside all disputes, and work in combination with the men in the trenches'.[64]

Come the spring of 1916 and Dalton was back in the trenches and tunnels of northern France. Then after a short home leave, he returned to active duty, transferring to the late-formed 258TC of the Royal Engineers. Enoch was now adapting his subterranean skills near to the French town of Loos-en-Gohelle, not far from Lens, an important coalmining region. The progress of 'silent work' underground in early August was interrupted by mines exploded by the Germans and subsequent bombardments of mortars and grenades, according to the War Diary.[67] After a lull, at 4.30am on 14 August, the aftermath the explosion of a large enemy mine resulted in the deaths of Dalton and four others, killed 'by a rush of gas'. Thirty-five years old, a letter sent by Enoch's commanding officer to his widow, Ann, suggests what may have happened.[68] Initially, three tunnellers were overcome by poisonous gas following the blast. Enoch, along with another volunteer went down a mine shaft and into the workings in a highly dangerous rescue mission but both were overcome by the deadly fumes.

Many soldiers composed 'posthumous' letters to be sent to loved ones in case of death, and Dalton's moving message to his wife, Ann, has survived, concluding with:

> . . . I have gone to the 'Great Beyond' there to await you and those who we love. May Dear Memories comfort you all, and Love abide with you. May your sorrow be light and your tears few, and your days long and bright. May our children be all that Love may Desire to you, and see that you are the Guide and Philosopher of their future days. In Life and Death always the same devoted Lover and Father.

Enoch's unselfish actions during 1915-16 were exceptional acts of bravery by any measure but unlikely to have been seen so by the man himself. It is nigh impossible to study Dalton's papers without getting a feel for his great empathy for his fellows, irrespective of nationality.

Another moving letter, written on 2 June 1916 by Dalton to the widow of Welsh miner Eli Davies survives among the Dalton family papers. It sums up how much he bonded with the men in his command and their families. Davies died after becoming overcome by gas following a mine explosion in the Hohenzollern Redoubt area on 21 May. Sending her a notebook belonging to her late husband, Dalton informed her that Second Corporal (Harry Russelbury) Wenlock DCM was alongside Eli at the time, losing his own life in a rescue attempt. Enoch assured her that they were buried 'side-by-side' in the Town Cemetery at Noeux-les-Mines, Bethune.[69]

Enoch Dalton himself was laid to rest in St Patrick's Cemetery, Loos-en-Gohelle, not too far from his old billet at Bethune. His gravestone (CWGC : 1.A.2.) is inscribed with a most fitting tribute: GREATER LOVE HATH NO MAN THAN THIS THAN TO LAY DOWN HIS ARMS FOR HIS PAL.

Enoch's descendants are extremely proud of his achievements and perhaps even more so for his modesty, humanity and great love of his family. And for all of us, his story is a wonderful testimony of what it was like for an ordinary, somewhat reluctant 'miner-soldier' serving as a tunneller in unimaginable conditions on the Western Front during the Great War.

Frank Depperriaz Perry

Sapper Perry's unusual middle name, maybe an indication of a Continental family heritage, must have made him stand out on official forms and lists, though his main military papers have not survived. However, thanks to the research and writings of local historian Mick Manise, we have an account of his short life.[70] Perry's formative years were spent in the north Warwickshire village of Shuttington, and after his marriage to Hetty in 1909, he lived just across the Staffordshire border at Amington, near Tamworth. The son of a coalminer, Frank followed his father's footsteps, working as a hewer at Pooley Hall Colliery, Warwickshire, at Polesworth. He must have escaped from the pit for a while, volunteering for military service, as he was called up from the Reserve immediately after war was declared in 1914, aged thirty-one; and by 12 August he had already disembarked in France, part of the 2nd Battalion of the South Staffordshires.

Like Dalton, Perry was one of the transferees who helped form the first tunnelling company of the Royal Engineers, selected because of their special mining skills. He experiences with the 170TC include some of the worst horrors of tunnelling in the Givenchy-Cuinchy sector, as described by Dalton, Blunden and Graves. On 20 August 1915, whilst Dalton was recuperating due to rheumatism, Perry and three of his comrades were killed when the Germans exploded a mine under a new gallery that they were constructing. He was interred in a shared grave alongside two of the sappers who died with him, H. Lee and M. Howell, in Cambrin Military Cemetery (aka Cambrin Chateau Cemetery), east of Bethune, on the La Basse Road, a few hundred yards from the front line. Next to Perry is the grave of Sapper F.J. Webster, a neighbour of Frank, who lost his life in similar circumstances and in the same area only a week previously. Perry's name is one of thirty-two 'of the fallen' inscribed on a 'pillar of remembrance' memorial dedicated to Pooley Hall miners who lost their lives in the Great War, sited near to their old colliery.

George Thomas Middlemiss

Durham miners were always regarded as tough, ideal enlistees for the new tunnelling companies. George Middlemiss, aged twenty-seven, 5 feet 6 inches tall and weighing 144 lbs, an experienced 'faceman' from Marley Hill Colliery near Gateshead, was 'specially enlisted' for the 175TC. Although having no military background, George was hurriedly 'fitted-out' at Chatham and rushed to the Front within a month of signing his attestation form on 1

Sapper George Middlemiss's medal entitlement includes the 1914-15 Star, shown here. Brian Elliott

June, mustering as a 'tunneller's mate' at 2s 2d a day. His pay was a lot less than the 'six-bob-a-day' demanded by Norton Griffiths for his 'moles', and also way below the average daily earnings of 6s 5d of all miners.[71] George was recognised as a fully qualified tunneller on 1 August 1915, presumably getting paid accordingly. But, as can be seen below, it was an apprenticeship that was both dangerous and spectacular.

Middlemiss was immediately deployed in underground warfare, in the Hooge Sector, the 175TC – under Major J. Hunter Cowen – positioned close to the Menin Road (at the same time as Leeming was directing operations at Cuinchy). He was part of 200 new miner recruits assigned to Cowan, enabling offensive mining to start in earnest. Direction was actually delegated to a young officer from Barnsley, 21-year-old Captain Geoffrey Rowley Cassels. Cowan and Cassells accepted a challenge from Brigadier-General A.R. Hoskins – after several apricot brandies – to tunnel under the German defences and blow up 'Chateau Hooge', the adapted stronghold held by the Germans. Cassels was a mining engineer but exaggerated the amount of explosives required, though in all fairness the principal use of ammonal was experimental, never having been used before in a military context. After shaft sinking and gallery driving in difficult conditions George and his mates managed to deploy twenty-four detonators, linked to 4,500lb of explosive. Blown at 7pm on 19 July, it was only the second (after Hill 60, on 17 April) British underground offensive in the Salient. The huge blast ignited a German ammunition store, pieces of buildings and an uprooted tree hurled into the air as though from an erupting volcano, with the macabre view of body parts of Germans raining to the ground. 'Cassels' Crater' reflected the immensity of the impact: 6 metres deep and 40 metres across. Several hundred Germans were killed in their redoubts but unfortunately ten 4th Middlesex men in a forward position were also buried in the debris.

Cassels expected at the very least a reprimand but was congratulated in person for his achievement by none other than Lieutenant-General Sir Edmund Allenby; and shortly afterwards he learnt that he would be awarded the MC, his 'friendly-fire' killing of the Middlesex soldiers 'forgotten'. Such was the to and fro of mutual offensives that within eleven days the Germans had regained this small patch on the Western Front, activating a new weapon of destruction in the process – the flame-thrower. The 175TC's and Cassels' involvement is described by military historians Alexandre Barrie's pioneering work *War Underground* and more recently in Tony Bridgland and Anne Morgan's biography of Norton-Griffiths.[72] Peter Barton et al's book *Beneath Flanders Fields* also contains a graphic account of 'the big blow' at Hooge.[73]

The 175TC had moved to Hill 60 in August 1915, driving strategic tunnels from behind the British lines towards and under German positions; and over the next couple of years functioned as a somewhat itinerant unit, moving from battlefront to battlefront in complex interchanges of duty. If ever there was a reminder of the danger of mining it became patently evident in O-Sector (Neuville St Vaast) on 21 May 1916 when nine '175s' lost their lives, overcome by gas, their bodies found at the shaft bottom and in associated galleries by 'Protomen': trained mine rescuers wearing self-contained breathing apparatus.

Middlemiss was almost certainly a part of the preparatory work prior to what has been described as the 'largest mining attack in the history of warfare', at Messines Ridge, on 7 June 1917.[74] The Sapper's final few months as a tunneller was with the much-travelled 254TC, interrupted in August of 1918 when he was hospitalised in France after fracturing his left leg in an apparent accident. The previous year, after Messines, he had continued his mining activities with the 175s, most notably at Vimy Ridge. It was an extraordinary three years and two months of front line subterranean work for the slightly-built Durham miner.

George Auty Chatt

Aged forty-one, Chatt was part of several drafts of veteran miners who joined the 177TC at Terdegahem or St Sylvestre Cappel in May and early June 1915, prior to setting up headquarters just west of Chateau Couvie on the Poperinge to Proven Road. Here, they were primed for front-line action fifteen miles away. Iain McHenry's *Subterranean Sappers* provides much detail about the company.[75] From Barnard Castle, County Durham, Chatt would have come into contact with other Geordies, perhaps 43-year-old John Bradley (Bishop Aukland) and George Campbell (Southmoor), who at forty-nine was one of the oldest of the recruits. By 18 June, the unit had a complement of 243 officers and men.

Shifts of tunnellers from the newly formed company were soon deployed in the Ypres Salient at Fly Farm, sinking a pair of shafts and driving galleries from the British lines below and just beyond the German trenches. The age-old problem in mining – water seepages and cave-ins – hampered progress. Incessant rain meant dealing with loose, unstable sand; and in the background was the ever-present threat of enemy shells. What a baptism of fire for George and his tunneller mates.

Between 25-27 July, 177TC took over the Railway Wood sector, the euphemistic 'bump in the landscape' on the Bellewaerde Ridge near Zillebeke, that was strategically important for occupiers, allowing opposition trenches and positions to be overlooked; and it was here 'where 177 and its miners would really cut their teeth'.[76]

Sadly, it was the 'miner-vets' who suffered the most. The required physicality for work here was extreme and unbearable at times, even for the younger, fitter men; and the mental strain that had to endured must have been hugely debilitating. Concentration, however, was a matter of life and death during the 'never ending' spells of listening duty, uncomfortably sat or knelt in dim candle light for five hours or more at a time. Fatalities began to accrue. Sapper Robert Onions, a 26-year-old former Staffordshire miner, was one of the first, losing his life whilst working on the surface above an extended gallery on 2 September.[77] Twenty days later, two of George's Geordie compatriots, Henry Newby (from Esh Winning) and Arthur Thomas (from Tow Law), were killed by German shellfire.[78]

The first offensive mine blown by the unit took place at 4.19am on 24 September, just two days after the deaths of Newby and Thomas. It created a 25-foot deep crater, 110 feet long and 85 feet wide. A German redoubt and 80-100 Germans were believed to have been killed. Typically, a counter attack pushed the British positions back to their previous lines, heavy infantry casualties ensuing.

Chatt's active service with the 177TC continued in the most dangerous of contexts from late September, during the month-long Battle of Loos. Perhaps he remained busy amongst the warren of underground workings at Railway Wood, or maybe amid the dugouts at Ypres or Yser Canal machine gun positions. If ever a reminder was needed of the fragility of life in the Railway Wood sector, the Ox and Bucks lost two officers and thirteen men as a consequence of a German blow on 17 October, a huge crater created damaging their trenches. The face-to-face horror of underground warfare – and its consequences – emerged on 28 October when German Pioneers broke into Working 5, triggering an underground shoot-out and the charging and blowing of a mine that effectively destroyed the workings.

On 14 December, after the Germans had exploded a mine creating yet another large crater, Chatt lost his life whilst on listening duty. Buried by many tons of debris, his body was never recovered. The shockwave from the blast was so immense that an infantryman was killed on the surface.[79]

Sapper George Auty Chatt is remembered on the RE Memorial on the hilltop at Railway Wood which commemorates twelve soldiers (eight from

The Railway Wood war 'cross of sacrifice' memorial is located close to Zillebeke and numerous mine craters which can still be seen. Brian Elliott

The RE's monument marks the place where twelve Royal Engineers (eight from 177TC) and four attached infantrymen were buried, killed whilst tunnelling during the defence of Ypres between 1915 and 1917. Brian Elliott

Sapper G.A. Chatt's name is inscribed on the same panel as his 177TC's colleagues Carter (KIA 13.6.1916) and Cotterill (KIA 22.7.17). Brian Elliott

177TC) 'killed and buried in action below here' during mining operations during 1915-17. Railway Wood is the smallest but one of the most emotive of the Great War memorial sites in the fields of Flanders.[80]

Welsh miner-tunnellers

Ritchie Wood's *Miners at War 1914-1919*[81] shows that South Wales miners were often at the forefront of tunnelling operations, working alongside pitmen from other British coalfields. Wood's meticulous research uncovered a total of 388 individual Welsh miners distributed throughout the twenty-five tunnelling companies, varying from forty-two (in 171TC) to just four (256TC), with fifty others not allocated to a specific company. Ritchie's narrative is especially compelling because of his inclusion of five of the TC's as case studies (170TC, 171TC, 172TC, 253TC and 254TC), covering the formative (1914-15) and later (1916-18) periods of each unit's activity; and also because of his frequent citing of the names, backgrounds and circumstances of individual miners. Overall, Wood's academic thesis and book are based upon his Register, a remarkably informative and unique database containing details of 438 south Wales miners serving in the tunnelling companies.[82] Ritchie has also identified 207 south Wales miner-tunnellers who died whilst on active service, about 15% (1,421) of all known British tunneller fatalities.

Sergeant James Herbert Spencer, 1st Battalion Monmouthshire Regiment and 171 Tunnelling Company RE

The first tunnellers faced a most extreme, unsafe form of mining, subjected to risks way beyond what they would have had to face at their home collieries. One of the most dangerous things to deal with when working in cramped workings was poisonous gas. Unfortunately, the use of breathing apparatus to facilitate rescue was either absent or deployed with limited practice and training at this time. The case of Sergeant James 'Jim' Spencer is fairly typical.

Spencer was a 26-year-old Monmouthshire miner, living with his parents at Cefn Road, Blackwood. As a local Territorial, he sailed to France with the 1st Monmouths on 14 February 1915. At the time of his death, less than three months later, he was attached to 171TC. Selected or volunteered, Spencer was one of forty miners from the 1st Monmouths known to have transferred, probably before embarkation. The circumstances concerning Spencer's death at Hill 60 were as follows:

> The Germans blew a camouflet on 2 June 1915, under the end of a British gallery and succeeded in burying one man and half burying another. Rescuers were aware there was a lot of gas in the area. In

The 'war-created' landscape at Hill 60 attracts many visitors, perhaps most poignantly from those whose recent forebears fought and died on or near the historic site. Brian Elliott

attempting to rescue one of the men, Spencer was overcome by gas and remained in the gallery. Lieutenant Kinloch, wearing Proto apparatus, managed to tie a rope around Spencer but accidentally knocked off his mouthpiece and on being overcome by gas went out of the shaft. Spencer was later pulled out but was found to be dead on reaching the surface. He was buried behind the trenches.[83]

Spencer's grave was subsequently lost but he is commemorated on the Menin Gate Memorial, Ypres (Panel 50) and, locally via a memorial inscription on a church bell in St Margaret's, Blackwood. Second Lieutenant James M.T. Kinloch, the RE officer who had tried to save him, was himself killed along with two 'other ranks' when the Germans exploded a large mine a few weeks later, on 11 July 1915.

Sapper William Arthur Lloyd, 179 Tunnelling Company

Family and local historians, military historians and specialist archaeologists continue to discover new information about the underground war on the Western Front. In 2016, bound for Lochnagar Crater, I was able to divert to the small but well preserved war landscape on and around the 'Glory Hole', during an open day event and exhibition.[84] Here, the La Boisselle Study Group (LBSG) have been able to uncover much about the secret subterranean war at this unique place, combining findings from careful excavations with a range of sources and resources, from previously unseen archives to DNA tests.[85]

One of the most spectacular examples, reported in the media in recent years, concerns a former north Wales miner who was killed in action at Ovillers-la-Boisselle during preparations for the Somme Offensive, a few days before Christmas 1915. William Arthur Lloyd, an able and experienced 37-year-old pitman from New Broughton Colliery, near Wrexham, was a welcome addition to the 179TCT, operating in the southern sector of the Somme. It was a terrible posting. In November 1915, six men of the 179TC and seven attached infantrymen were killed by a German mine. Six weeks later, on 19 December, William was at work with four mates in a low tunnel in the chalk rock below No Man's Land, advancing to prepare the way for laying explosives. German counter tunnellers, after hearing their presence through a microphone, detonated a huge mine that killed the group, and in so doing creating a 40-metre crater on the surface.

Sapper William Arthur Lloyd. Brian Elliott

In 2013, thanks to the archaeological explorations and research of Peter Barton, his LBSG colleagues and historian Simon Jones, Lloyd's great granddaughter Lesley Woodbridge, was not only able to make a pilgrimage from Staffordshire to the spot at the Somme battlefield where he died but

The much-disturbed crater landscape at La Boiselle is now protected because of its historic importance. Here an exhibition guide points out interesting features to members of a WW1 tour group, enjoying an unexpected visit. Brian Elliott

A goat rests on the crest of a huge mine crater at La Boiselle, its eyes directed across the abyss towards a passing battle tour group. Brian Elliott

Entrance to an exploratory adit (tunnel) at La Boisselle. Brian Elliott

A 'tunneller's kit' on display at La Boiselle. Note the short handled excavation spade. Brian Elliott

descend underground and reach the blocked site where he was working, an experience described as 'a historical first' by Barton.[86] William left a widow, Elizabeth and six children and is commemorated on the Thiepval Memorial to the Missing of the Somme (Pier and Face 8A & D).

* * *

Authenticated stories about our coalmining ancestors, collected and curated by their descendants remain one of our most important sources for understanding the generational impact of the Great War. Below are just two examples, based on family history research.

Senghenydd veterans

William 'Will' Fisher

Will was near the end of his night shift at Universal Colliery when a massive explosion ripped through the workings on 14 October 1913, just as the day-shift men had started. A total of 439 men and boys lost their lives. Fisher was one the lead rescuers, searching for survivors in very dangerous conditions. One of his colleagues was killed in this task, and he almost lost his own life. Thirteen months later, in January 1915, Fisher was welcomed into the Royal Engineers, initially as a Sapper with No.2 (Holyhead) Siege Company (Royal Anglesey RE). We know much about Will through his surviving letters and diary entries reproduced by his niece as *Requiem for Will* (1997).[87] Here's a typical entry, for 6 September 1915:

> Been three days at the ramparts, south-east side of Ypres. Repairing trenches, in charge of a small party of chaps, a COPORAL'S job. The rain these last few days has considerably increased the hardships of campaigning, the trenches being in such a state with mud and water and, to make matters worse, the lack of facilities for drying one's clothes in camp. The dugout roofs let in water too. So far wet feet and damp clothes have not caused any ill effects, but I'm afraid they will ...

Will Fisher. Elaine H. Fisher

Will's worst fears materialised. By 24 September he was at No.10 Casualty Clearing Station, Poperinge, very sick with tuberculosis but he was put to work working at his convalescence camp within a few days and – now as Lance Corporal – was back digging dugouts on 10 November. His excavation and fortification experiences were extensive throughout the war, working alongside tunnellers, even though he was unable to officially transfer to a specific company. Latterly as Company Sergeant Major, Fisher served an extra year in the army until discharge in March 1920 but sadly succumbed to the disease he had concealed when joining up in 1915.

Alfred 'Alf' Gordon

Aged only sixteen, Alf had finished his night shift on 12-13 October 1913 at 7am and was bathing at his Abertridwr home when 'commotion' all around him indicated that a big explosion had occurred at his pit, the Universal. In a state of shock, he rushed through the streets towards the pithead, amid the shouts, sobs and screams of his neighbours. The Aber Valley mountains that he had got to know were 'black with men from neighbouring collieries'. It was a terrible disaster with unimaginable consequences for hundreds of grieving families, now pushed into greater poverty. Less than a year later, Alf had found work with his uncle, albeit in very wet conditions at the Windsor pit. After witnessing patriotic scenes in the garrison town of Monmouth, when Kitchener volunteers were given a grand send off, on 2 September 1914, Alf and a few of his mates went to Caerphilly 'to sign up'. He was welcomed into the Prince Albert's Somerset Light Infantry, although at the age of seventeen years and four months he was an underage recruit and well under the minimum age for overseas posting when dispatched to France with the 7th battalion in March 1915. It was the start of an extraordinary but in many ways typical war experience for thousands of young ex-miners, his recollections recorded by his grandson David Barnes, whose book *Black Mountains* is an account of a young Welsh miner's war.[88]

Promoted to Corporal whilst still seventeen, Alf was soon in the trenches, occupying the front line and making excursions through No

Alfred 'Alf' Gordon.
David Barnes, *Black Mountains* (2002, Y Lolfa)

Man's Land, including leading a scouting mission to an abandoned German position. 'OTT' (over the top) formed a frequent part of his onward narrative, often crawling on his stomach amid the corpses of British, French and Germans.

Alf described one of his most memorable OTT experiences, during the spring of 1915 when trying to deploy respirators that were useless against waves of German gas attacks:

> Not for the first time we felt vulnerable and exposed. The slaughter was terrible. The enemy moved up behind the gas and let loose with machine gun fire. Our platoon sergeant was killed, and I led the remnant of the platoon back to the trenches we had vacated . . .The chlorine had got into my lungs and I felt dreadful. That was how I spent my eighteenth birthday.[89]

Later, when Alf was on home leave on account of his gas debility, he stripped off his louse-infested clothes only to see his mother break down in tears at the sight of his emaciated body.

Back on the devastated Somme battlefield, and now a young platoon sergeant, Alf experienced some of the worst scenes imaginable, one of the saddest being when his much-admired C Company commander 'Captain Hatt' (Edward Beach Hatt) was fatally hit by a German shell when running between trenches, only a few yards in front of Alf, near Trones Wood, in August 1916.

Gordon felt 'a surge of new optimism' on the Somme when he saw tanks for the first time, at Flers Courcelette on 15 September 1916, but he had a personal setback a day later when a machine-gun bullet penetrated his

Tanks were used for the first time at Flers Courcelette. Taylor Library

right arm, necessitating hospitalisation and a longer spell of enforced leave. Alf's active war was curtailed but not ended as he finished up as an anti-gas instructor back at Rollestone camp on Salisbury Plain where he had trained as a rookie soldier.

Discharged from the army and married to Dorothy, Alf was back down Windsor Colliery in 1919, albeit only briefly at his old, very strenuous job which was far from suitable given his parlous state of health. But forms of pit work continued, Alf devoting his time to the health and safety of miners, becoming so competent that he rose to become the Area Safety Engineer for the Rhymney district. Alf Wood, Senghenydd and Somme veteran, died peacefully, aged seventy-three, a remarkable survivor and a huge credit to the mining industry of Britain.[90]

Mining families 'at war'

Most coalminers during the Great War lived in close-knit communities, employment provided by usually one, two or even more collieries if they lived within a 'commutable' journey from home. Households were often large, even cramped properties finding attic or make-shift space for a relative or lodger. The incomers were usually new, distant arrivals, attracted there by the prospect of pit work. Within such contexts it was no surprise that mining brothers and relatives left their beds and jobs to enlist, ultimately for service abroad. Very often younger brothers followed the military footsteps of their seniors, escaping from one danger (pit-work) to another (action on the Western Front).

Online searches of Great War period local and regional newspapers in coalfield areas reveal many examples of sibling military service of former miners during 1914-15, invariably reported in patriotic terms, perfect examples for others to follow. Increasingly intermingled with these, are cases where one or more brothers were missing or known to be killed in action, situations that became more commonplace in 1916 and continued through to the Armistice and beyond; and though 'gallantry' reportage may have countered this, it could not hide the gloom and heartache of many families who suffered multiple bereavements, loved ones lost on active service. Repeated family names listed on coalfield area war memorials bear stark testimony to familial sacrifice.

A Warwickshire miner-soldier family: The Launchburys [91]

The Launchbury's were in many ways typical of the immense contribution that larger mining families made during the Great War. Four brothers 'wore the uniform' as well as three brothers-in-law, several nephews and grandsons.[92] A colliery labourer, Alfred Launchbery and his wife Sarah Ann had eight children in residence with them according to the 1901 census, at Grange Lane, in the ecclesiastical parish of St Laurence with Longford, near

Gunner Mark Launchbury and his British War and Victory medals. Brian Elliott

Coventry. The four males are named as Joseph (aged twenty-three, a hewer), Alfred (seventeen, a horse [pony] driver), William Charles (eleven) and Mark Stanley (four). Their eldest son, also named Alfred, aged twenty-eight and another, John, twenty-five, lived in separate households.

By 1911, the youngest son, Mark Stanley, now fourteen, had joined one or more of his brothers working underground as a 'motor driver'. Although several mines were within a short distance of home, the Launchbury's had a close association with Exhill Colliery. In the early hours of 21 September 1915, at a time when several of the brothers were in military service, fourteen miners at this small 'family pit' (375 at work on 'disaster-day') suffocated following a freak underground fire. A paraffin lamp was accidentally dropped by a workman during a routine maintenance operation, igniting the woodwork of the headgear above the downcast shaft, deadly fumes and smoke filling the tunnels below. Due to absence or good fortune, one or more of the Launchbury family may have escaped death or injury in the pit, but also faced similar danger whilst on active service.

Bad news reached Alfred and Sarah's Longford home concerning their second youngest son, William Charles, a regular soldier serving with the 1st Coldstream Guards, part of the BEF from 7 August 1914, who was reported missing. William, unmarried, had in fact been killed in action on 25 August during the retreat from Mons. More heartache followed. Alfred junior, at thirty-two the oldest of the brothers, a former Exhall miner and a Colour-Sergeant in the 3rd Coldstream Guards, with thirteen years of service prior to embarkation to France and Flanders with the BEF on 12 August 1914, lost his life near Vermellen, on 27 September 1915, during the Battle of Loos – less than a week after the Exhill tragedy. Alfred left a wife and three children. Better tidings arrived regarding two other of the brothers, Lance-Corporal John Launchbury (7th South Wales Borderers) and Gunner Mark Stanley Launchbury (Royal Field Artillery). With three sons-in-law and two grandsons now also serving, it must have remained a very anxious time for

Alfred and Sarah, not helped when yet more terrible news came through concerning the death of another nephew, 'killed in the Dardanelles'.[93]

By 1917, Sarah was in a parlous state, not only reeling under all the distress but also struggling to make ends meet, her husband now having passed away. A benefit concert by a Pierrot Troupe raised £32 for the family. Some redress came, perhaps, when sons John and Mark Stanley actually came home after the war, and were now free to work in mining.

Private Alfred Watkin, 13th Northumberland Fusiliers

Like Alfred Launchbury, Alfred Watkin also lost his life during the Battle of Loos. He was, initially, reported missing and 'wounded in the face' following 'a heavy assault on Hill 70'.[94] Subsequently, Watkin was recorded as killed in action, on 26 September 1915. Watching through his field glasses, Robert Graves in his book *Goodbye to All That*, referred to the attack as 'a real breakthrough' and then went on describe the unenviable task of collecting what dead and wounded they could, acknowledging that the Germans 'behaved generously' by not firing during the retrievals; whilst the trenches 'stank with a gas-blood-lyddite-latrine smell'.[95]

Alfred Watkins and his British War Medal. Brian Elliott

The War Diary is quite explicit regarding the 13th on 25-26th, its commanding officer describing 'another attack . . . but unsuccessfully and the remnants of the force . . ' withdrawing 'to the German trenches captured during the morning'. The casualties, those known to be killed or 'missing believed to be killed' were immense: 395 'all ranks . . . approximately'.[96] Alfred's widow, Nellie, looking after infant daughter Amelia, was unable to get any compensation for many months due to the inexcusable delay in getting official confirmation from the War Office of her husband's death. One of Alfred's younger brothers, Fred, survived the Somme, returned to mining, but sadly died at the age of thirty-eight. Private Alfred Watkin is commemorated on the Loos Memorial (Panel 20-22), on a family grave in Wombwell Cemetery (UC.2.2591) as well as on public war memorials in Wombwell and Hemingfield, near Barnsley.

Alfred Watkins was born and brought up in Lundhill Row, a long, distinctive community of miners' cottages at the edge of Wombwell, not far from Cortonwood Colliery, and within walking distance of several other mines. His father, Joseph ('Joe'), who had married Elizabeth, a 'Staffordshire lass', had swapped relatively poorly-paid textile work in west Yorkshire for better miners' pay, rising to become a 'contractor', a self-employed miner in charge of a small team of men and boys working and servicing a coalface. 'Piece-work' wages would have been shared out by Joe, to his team, each week in the pit yard. Joe and Elizabeth had many mouths to feed, the 1901 census showing the presence of seven children, aged four months to thirteen years old; plus two teenagers already working underground, probably as part of Joseph's 'gang'. The oldest child, Alfred, aged thirteen had yet to start work. Thus there was a household of ten in what was probably a two or at the most a three-bedroomed property.

By 1911, Alfred was living with his soon-to-be in-laws and new wife Nellie in a more spacious property nearby. Married in the same year, his trip to Sheffield to enlist in the Northumberland Fusiliers was probably a more than an impulsive decision given his background, an opportunity to 'see the world' but soon to be back home with his new baby daughter. His time on active service in France was probably little more than a few months.

Many hundreds of former miners participated in actions at Loos, the first large-scale British offensive of the War, where poisonous (chlorine) gas was deployed for the first time. Alfred Watkin was one of over 20,000 casualties suffered by 28 September; whilst Arthur Launchbury remains one of the 59,247 losses accumulated in the main and subsidiary attacks.

***

Fife miner Private Robert Anderson Dunsire, 13/Royal Scots, was awarded the VC for his bravery on the Hill 70 attack, on the same day that Watkin lost his life. Later, at Haines, on 8 October, Lance Sergeant Oliver Brooks, 3/Coldstream Guards, a former 'carting boy' at Norton Hill Colliery, Somerset, also got a VC following his extraordinary action and leadership in the recapture of a section of British trench.

***

Royal Naval Division 'miner-recruits'

Although most miners joined infantry regiments, some enlisted in the Navy, even from coalfield areas many miles away from the sea. The coal town of Barnsley, famous for its raising of two Pals regiments, mostly of former miners, had forty men enlisted with the Plymouth Battalion Royal Marines Light Infantry, involved in action during the Gallipoli campaign. The survival rate of these volunteers was low: sixteen killed in action (or fatally injured), two others made prisoners of war, one man classed as a deserter, another became an officer; and for nine others their service was ended early due to invalidity. Only eleven of the forty served until then end of war and demobilisation.[97]

A boy 'miner-sailor': Louis Hodgkiss

'A Sailor Lad': Detail from a studio photograph of Louis Hodgkiss. Winifred Hodgkiss

About one in three naval recruits were teenagers, well under military combat age. From Wigan, Louis Hodgkiss went down Walthew House Pit for the first time on 2 December 1914, his fourteenth birthday. Initially rejected at his local army recruitment office on grounds of his appearance and height (he was only 4 feet 10 inches), he was accepted into the Royal Navy less than two years later.

Louis's military papers record his first service date as 3 October 1916, when he was still under the age of sixteen, a 'boy sailor'. A memoir about his life, written by his second wife, the writer and academic Winifred Haward Hodgkiss, states, however, that his service was a few months earlier, on board HMS *Barham*, during the biggest sea battle of the war, Jutland (31 May-1 June 1916).[98] His naval records do confirm service on the *Barham*, though not until 1917. Whatever his starting date was, Hodgkiss was with the RN right through to the Armistice, when he was still only eighteen.

Louis Hodgkiss remained a stoker until his discharge in 1922. His leaving was a far from honourable affair, as he accumulated a long list of charges in order to curtail his service. A most bizarre misdemeanour occurred when on shore leave he absconded from duty and got set on at Swindon Colliery under an assumed name. The resultant 'desertion' charge resulted in a 'let off' 90-day detention, though the extent of this is not entirely clear on his official record.

After a period as a stoker on merchant ships, Hodgkiss returned to the mines, at Moston pit, near Manchester, quite rapidly advancing to become a coalface man. In a far from harmonious marriage, he began writing about mining, had the sense and confidence to get an agent. The first outcome, was when his manuscript 'Naked Men' (Lancashire miners working in hot conditions, wearing little more than clogs or pit boots) was accepted for publication under the title *The Black God's Shrine* (1935). Numerous other stories and plays appeared in quick succession, for example *Deep Shafts* (1936), *Underground Legion* (1937), *The Deputy* (1938) and *Trapped Below* (1939), each very well received by the regional press as well as *The Stage*. All were penned whilst he worked as a collier. Almost crushed to death in a roof fall, some of Louis's drama work was featured in the acclaimed Ealing Studio film *The Proud Valley* (1940), starring Paul Robeson, and set in south Wales. After his divorce, Hodgkiss married his long-term partner Winifred Haward Hodgkiss (who earlier had changed her surname to Hodgkiss by deed poll) in 1972. Louis Hodgkiss died at Skipton, Yorkshire, pretty well 'unknown', six years later. His own life story is a quite remarkable journey, from boy miner and able seaman to collier and author-playwright, a somewhat forgotten figure but for his second wife's memoir.[99]

A 'miner-Marine': William James Jones

It was with some relief when Mr and Mrs William and Hannah Jones, at Christmas in 1915, received a greeting card from the eldest of their seven children, Gunner William 'Bill' James Jones, who had 'escaped the pit' to join the Royal Marines in 1914 at the age of seventeen. William senior worked as a hewer at Wharncliffe Woodmoor Colliery, whose muckstack dominated the skyline at the end of their street, Grays Road, at Carlton, near Barnsley, where I spent my boyhood. Known as the 'Jones Singers', the family were well-known for their musical abilities, many villagers cramming into their 'best room' to join impromptu sing-songs. Bill junior wrote home from Eastney Barracks, Portsmouth, which must have seemed a million miles away.

Jones went on to have an eventful Navy career, on board (among others) the battlecruiser HMS *Inflexible*, according to the memories of his youngest brother Lloyd, who provided details to his local newspaper many years later.[100] As a gunner, Bill experienced some of the most noted sea battles of the war, including Jutland in 1916; though his ultimate survival was, according to Lloyd, because of two spells of enforced shore leave, due to serious illness, at times when many of his mates lost their lives. On demob, Bill returned to the pit where he had started work alongside his father, as boy of fourteen. He soon left the mine, for a much cleaner and safer career, as an 'insurance man', and lived until the age of eighty-six.

Gunner William James Jones.
Barnsley Chronicle

The battlecruiser HMS *Inflexible*. Public Domain

Chapter Four

SILVERWOOD: A WARTIME COLLIERY

The story of Silverwood Colliery during the First World War was similar to that of many of the large coal mines in Britain. Miners rushed to enlist in large numbers and many of the households that served the pit lost one or more members. This great escape from the dark was, as we have already seen, often a case of swapping one dangerous job for an even more life-threatening one. The teams of miners 'left behind' were

A small group of miners (and a local 'paperboy') in front of 'Dalton Main' coal wagons, c.1910.
Brian Elliott

A model colliery complex

By 1914, only nine years after its first coal was extracted for sale, Dalton Main Colliery, near Rotherham in South Yorkshire, had become one of the largest coal mines in the north of England and was regarded as the biggest single-seam colliery in Britain.[1] Its exclusive use of the famous Barnsley Seam was reflected in the second word of its original name: 'Main'; whilst its forename, 'Dalton', related to the local rural parish in which it was located, though many of its miners resided within the ancient parochial area of Thrybergh. Throughout the war, 'Silverwood' as it became known, traded as part of Dalton Main Collieries Company Ltd, which included the neighbouring but much smaller Roundwood Colliery.

Dalton Main's name was changed to Silverwood (after a small wood) in 1908-09, following a buy-out by the huge John Brown Company, when the mine's workforce was 3,108. This figure increased to 3,228 two years later and jumped to 4,295 in 1912. The smaller and shallower Roundwood Colliery employed 616 men and boys. Silverwood was connected to Roundwood, three miles distant, by underground headings and overland via a rail link. This arrangement enabled the latter's coal to be wound, processed and distributed from a new, larger and more modern site.[2]

Silverwood's status was such that it was chosen as a showpiece mine for the royal visit to south Yorkshire by King George V and Queen Mary in July 1912. The occasion, however, was overshadowed by the disaster at nearby Cadeby Main colliery when a total of 91 men and boys lost their lives in two explosions, the second one accounting for the death of many rescue workers.[3]

Location of Silverwood, Roundwood and adjacent collieries. Detail taken from 'Geological Map of the Yorkshire Coalfield'. Midland Institute of Engineers

Looking across Hollings Lane to Silverwood Colliery about the time of the Great War when the 311-acre site, with its distinctive lattice headgears, chimneys, and by-product plant had already become a large industrial complex. Brian Elliott

expected to produce the same or even more coal than before on depleted wages and in worsening working conditions.

Silverwood's owners supported enlistment, at least during the first two years of the War. This, despite the many practical and economic problems they faced due to an exodus of key workmen.

Whinney Hill miners' church

Strategically sited at the heart of the mining community, St Peter's function was that of a 'Mission Church', its Conventional District formed under the South Yorkshire Coalfields Extension scheme.[4] What a difficult job it must have been for Reverend Hubert Rouse Everson, the priest in charge in the war years, when news of multiple casualties from the Western Front and elsewhere reached Whinney Hill. The 'miners' church' did what it could to support and comfort local residents in dreadfully difficult circumstances, indeed its 'personal' toll included choir and Bible Class members, as well as numerous worshippers.

St Peter's Church, Whinney Hill. The church closed for worship in 2018-19. Brian Elliott

Volunteers and call-ups

Within a few days of the declaration of war, it was reported that 400 'Dalton coal' men had 'responded to mobilisation orders', the company's general manager, Mr J. Blenkinsop, assuring all of those accepted for service 'that their posts will be kept open to them'.[5] A few days later, the manager stated that 1,168 men from Silverwood and Roundwood were 'now serving' in the armed forces, an exceptional number from a single 'colliery'.[6] By 1917, under a 'Patriotic Silverwood' headline, it was reported that 1,181 Silverwood miners had in fact 'originally' enlisted.[7] According to the local miners' union, when their members assembled in January 1918, one-third of them had served their country.

Diversity and comradeship

The men who enlisted came from a variety of cultural backgrounds. A sample of fifty adjacent households recorded in the 1911 census for the Whinney Hill area is likely to be typical of the those throughout the mine's catchment area. Only 10% of the males classed as mineworkers were born locally, that is within the Rotherham area. The rest had migrated from as many as *fifteen* English counties, an extraordinary geographic spread, with 12% of the incomers arriving from as far away as London, Scotland, Wales and Ireland. However, many of the incomers made a fairly short journey from nearby mining areas in other parts of Yorkshire (41%), but there was a significant number (31%) that migrated from old mining areas in Lancashire, Derbyshire and the Black Country, notably Staffordshire. The mix of regional language and dialect of the incoming families must have been obvious to outsiders, and especially marked where the miners worked close to each other underground. Despite the great diversity, it was living in a 'pit village' – with its inherent comradeship and neighbourliness – that bound the community together. This commonality was one of the most underrated factors as to why so many miners were prepared to 'join-up', irrespective of their backgrounds.

A 'widow-maker' pit

The large collieries opened in the decade or so before 1914 experienced a variety of accidents that resulted in injuries and fatalities to their workforces. Silverwood was no exception. The day-to-day incidents, mainly roof falls and haulage mishaps, usually involved a single or small number of fatalities, but it was not long before the mine became known as a 'widow-maker'. At least 52 men and boys lost their lives since the sinking of the pit in the early 1900s through to the start of the war, an average of four deaths every year.

The war dead

It is very unlikely that we will ever know the absolute number of Silverwood and Roundwood miners who lost their lives as a consequence of the war (as against 'in the war'). A beacon-style war memorial, unveiled in 1923 and dedicated entirely to the miners from the Dalton Main company, was placed above Silverwood Quarries, overlooking the village and close to the colliery. On plaques, some 312 names were inscribed.[8] However, research continues to uncover examples of missing names; and there is no question that some former miner-soldiers passed away in the early post-war years, affected in various ways by their war service.

The 'Beacon Light' war memorial, c.1923. Andrew Featherstone

The war memorial was moved to its present site, by the parish Hall, in 1989, in order to allay vandalism, and the name plaques were placed within the building for safekeeping. Brian Elliott

Examples of recently discovered war memorial 'exclusions' include that of 23-year-old Seaman-Gunner J.E. Wright, who was reported as killed on board the destroyer HMS *Broke* on the night of 31 May 1916, during the Battle of Jutland.[9] Former Silverwood miner Private Charles Tyler was killed in action on 13 March 1915 whilst serving with the Royal Scots Fusiliers. Tyler, a Reservist with previous military service, had rejoined in 1914, his death and his photograph published in the town's newspaper.[10] That Tyler was residing in lodgings and his next of kin lived in Sheffield (where Tyler was born) may explain why he was overlooked when the memorial list was compiled.

Home Front and War Front

1914: *The rush to join the Colours*

Within days of the outbreak of war it was reported in the local press that 400 men employed at Silverwood Colliery 'have responded to the mobilisation orders', a substantial number from one pit in such a short period of time.[11]

Blenkinsop, the colliery's manager, assured the recruits that their jobs would be 'kept open'. As a further recruitment incentive, the Dalton company went so far as to offer a £2 bonus (about a week's wage) for any of its miners who enlisted.[12] A company-administered War Fund was also established, subscribed to by the 'officials, surface, underground, coke ovens and brickyard lads'. Its purpose was to provide financial support and assistance to the wives and dependants of those men who had enlisted. By November, 275 wives, 659 children and 73 'partial dependants' had benefitted from the £933 distributed from the scheme.[13]

On 14 November 1914, eighteen members of St Peter's Bible Class 'went in a body' to the army recruiting office and were accepted for service. Less than four years later, only one of them, Corporal Herbert Fisher, was reported as 'still fighting'. The others were either killed, discharged 'wounded' or 'taken prisoner'. Twice wounded, Fisher, who had joined up direct from Silverwood Colliery, got the Military Medal 'for conspicuous bravery on the Italian front' in 1918. He was twenty-one years old.[14]

It wasn't long before news of casualties started to arrive and within Thrybergh and district's close-knit communities the shock waves spread rapidly. A KOYLI recruit, Frederick 'Freddy' Briddon, 31-year-old son of a deceased Sheffield razor grinder, lost his life on 18 November towards the end of the (First) Battle of Ypres. Freddy was one of the most experienced and skilled miners who had joined the Colours, having worked as hewer in the Barnsley seam at the colliery. Listed just below the entry for Briddon on the 'Dalton' war memorial is another local name, that of Herbert John Bridgewater. A pony-driver in his youth, 28-year-old Rotherham-born Herbert, one of many of the local miners who joined the York and Lancasters, was another early casualty, killed in action on 5 October at Jouarre, in the Seine-et-Marne area of northern France.

News of the gallantry of former Silverwood men was usually reported in the local press but – as with casualties – might take months to appear in print. Seaman George Ripley had joined the Naval Brigade (more specifically the 'Collingwood' infantry battalion of the Royal Naval Division) only a month after war was declared. For 'conspicuous bravery' during the defence of Antwerp (3-9 October 1914), George became one the first ever recipients of the Distinguished Service Medal (DSM); and appears to have been one of only twenty-two of the Collingwood men to make it back to England from the Belgian trenches to their base at Crystal Palace. He was subsequently presented with an Illuminated Address by the

Seaman George Ripley wearing his DSM. *Rotherham Advertiser*

Thrybergh and Dalton parish councils in recognition of his achievements. Another local man, Private Charles Birchall, a stretcher-bearer with the East Yorkshire Regiment, was also 'address-honoured', though unable to attend the same ceremony. His Distinguished Conduct Medal (DCM), was awarded to him for the rescue of a wounded officer stranded between British and German lines.

By early December 1914, deep underground, twenty-year-old Philip Cunnane, from Dalton Brook, died before reaching hospital after a ton of rock had fallen on him during shot-firing operations, a reminder if ever there was a need for one, that mining coal was a very dangerous job for those men who continued to work.

1915: Wounded, missing and killed in action

The Dalton Main Company and the miners themselves continued to support those who had enlisted by providing financial help to their families. A voluntary levy of 2%, deducted from weekly wages of Silverwood and Roundwood employees, enabled regular weekly payments to assist wives and dependants. By late February it was reported that the sum of £2,300 had been paid out via the Dalton Institute at a rate of 3s (15p) a week to each wife and 1s (5p) for every dependent child; and 1,153 persons were now in receipt of 'relief'.[15] The dependant figure soared to 2,700 by October, as many more miners had volunteered.

The Dalton Main Coal Company also waived rents on their households where the head had joined the Colours, and allowed each house a free load of coal every six weeks. The company appears to have been coping well with sudden losses of key employees, and the loss of income from export markets due to the war. Another very practical problem was the fluctuating daily attendance of some of the remaining miners, making the deployment of labour very difficult. On 10 July, for instance, manager Blenkinsop stated that there were 300 fewer men at work on the Thursday compared with Wednesday.

When news filtered through to homes that a person had gone missing or was reported as injured, the resultant anxiety and distress was prolonged. It might be many months later when a named soldier or sailor was officially declared 'killed in action' or 'died of wounds'. The *Rotherham Advertiser* did what it could to get casualty updates, publishing individual appeals, but often to no avail. The case of a former Silverwood man, John Quinn, a thirty-year-old private in the Prince of Wales (West Yorkshire) Regiment, who had been called up as a Reservist in August 1914, is a typical. No letters from him had reached his wife Eliza

Private John Quinn.
Rotherham Advertiser

after 16 October, she and his friends 'exceedingly anxious' about his situation, especially when one returned letter from the Front was marked 'Killed in Action'; though another report 'cruelly' stated that he was still serving.[16] Confusion and heartache merged for many weeks until sad confirmation came through that John had indeed lost his life on 18 October 1914. A distraught Eliza now had three small children to look after on her own.

Cases where former Dalton miners were reported as 'wounded and hospitalised' were frequently mentioned, and of course some of these men died before getting back home. Private Arthur Hunt, a Reservist who had been on active service with the 2nd York and Lancasters for three years from 1902, was called into action early in August 1914, but by March 1915 was reported to be 'suffering from a gunshot wound and a fracture to the skull'.[17] Arthur's military record show that he had been taken to a French 'clearing hospital', operated on, and then conveyed to the military section of the London Hospital where he died of his wounds (though ultimately from meningitis), on 12 May 1915. Aged thirty, Hunt was 'buried by the military'. A telegram was sent to his already distressed widowed mother, Mary Ann.

Arthur Hunt, probably as a 'boy soldier'.
Rotherham Advertiser

Several Dalton miners were reported in the *Advertiser* as 'wounded' in 1915, though in reality there must have been very many more. These included Private Joseph Morris (KOYLI), said to have been wounded four times. In a letter to his wife, Morris described the 'storming of Hill 60' in some detail, his remarks including the following: 'We were ordered to take the German trenches in front of us, and we did so. The Germans ran like h___, but I am sorry to say we lost a good number of men and all our officers . . .'

One former Silverwood miner, Private Charles Sykes, serving in the 1st West Yorkshires, was reported as being a prisoner of war at Gustrow, in northern Germany. Enterprising 'Charlie' requested that his 'pit bottom' mates 'make a collection' – for himself! Fifteen shillings (about 75p) was duly forwarded to him by his mates, supplementing his meagre 'prison pay'. The Gustrow camp was a new and large facility, some 150 wooden barracks hurriedly erected in a pine wood.

About half of the miner-soldier fatalities from Dalton collieries in 1915 were under the age of twenty-five. A Northumberland Fusilier, Private Sydney Drabble was typical of the younger recruits, reported at the end of October as 'wounded at Hill 70', but he was actually killed in

Private Sydney Drabble.
Rotherham Advertiser

Silverwood: A Wartime Colliery • 89

Corporal Randell Gaskell.
Rotherham Advertiser

action on 26 September, and was only twenty-one when information was eventually conveyed to his mother, who lived in Thrybergh village. Sydney had previously worked in the Silverwood colliery offices.

War fatalities become more poignant and real where medals or memorabilia survive. Randell Gaskell (8th York and Lancasters) was killed in action on 23 September 1915.[18] Under twenty when he enlisted, Lancashire-born Randell had previously worked at Silverwood as a 'rope hand' (haulage worker), living at Whinney Hill. According to the letter sent to his father from his commanding officer, Gaskell died in a trench from shrapnel wounds when he was in charge of 'a guard of men', after a German shell 'exploded among them'. Corporal Gaskell was buried in the Sailly-sur-la-Lys Canadian Cemetery, Picardy.

Obverse view (left) of Gaskell's British War Medal, with its uncrowned profile of George V, who had visited Silverwood Colliery three years earlier. The reverse features St George on horseback, trampling over the eagle shield of the Central Powers and the emblems of death. The edge is inscribed: '14114 A. CPL R. GASKELL Y. & L.'. Randell's next of kin, his father, would have also been sent his 1914-15 Star and probably his Victory Medal; as well as a personalised Memorial Plaque ('death' or 'dead man's penny') and its parchment scroll.

At least forty Dalton miners died as a consequence of military combat during 1915. The majority, about three out of every four, had enlisted in local infantry regiments, most notably the York and Lancasters. A few of the casualties were from the Navy, also suggesting that for some, this service was preferred. A diversity of other regiments were evident due to 'diversions' (when the York and Lancaster Regiment was 'too full') or actual choices, again probably matching the pattern elsewhere. The regiments include the Royal Field Artillery, Northumberland Fusiliers, East Yorkshires, East Anglians and Royal Munsters. Interestingly, there were at least two recruits to the Royal Engineers (Archer and Downey), experienced workers, targeted for their mining experience and skills, who died serving the 180th Tunnelling Company.

In 1915, several 'accidental deaths' were reported of Silverwood miners who had lost their lives in the pit. These included two young pony-drivers: eighteen-year-old Harry Rackham, who was run over 'by a train of tubs', dying in hospital later, and Wilfred Bannister, nineteen, killed after being knocked down by his pony. Three experienced colliers also died: Thomas Stacey (39) and James Briggs (60), victims of roof falls; and John Haynes (46) died from pneumonia after being struck on his shin by a large piece of coal.

Sergeant William ('Bill') Ollivant, who had earlier seen service with the York and Lancasters as a boy soldier, sent home a few verses that he had composed, despite the inconvenience of a bombardment of German shells in the vicinity of his dugout. His poem concerns the carnage that occurred on 10 July involving his battalion, the 1/5 York and Lancasters, and includes the following lines:

> *The Huns started bombarding with grenade and shell,*
> *With explosions and flashes 'twas like being in hell;*
> *What with groans from the wounded and creeping o'er dead.*
> *'Twas marvellous how every lad kept a clear head.*[19]

The regimental war diary confirms this incident. Bill and his company were defending trenches taken from the Germans near the Yser Canal, northwest Belgium, when at least twenty-seven 'other ranks' and one officer were killed and 127 others wounded.

There is no doubt that by the end of the year the sad reality of war had become much more apparent in the parish of Thrybergh and the surrounding communities that served Silverwood and Roundwood collieries. As for 'poet Ollivant', he survived the war, even re-enlisting in the army in 1919, by which time he had found work in the steel industry rather than in mining.

1916: 'Dalton miners killed every week'

The Union flag appeared at half-mast above St Peter's for many weeks during 1916. It was a terrible period for the Dalton company's miner-soldiers and their families. The Somme Offensive accounted for at least twenty fatalities, most of them on that dreadful of all days: 1 July. The reported number of casualties during the year, approximately sixty, was probably well below the actual death toll. Once again it was the young men under the age of twenty-five, often newly married with young children, who lost their lives. Unimaginable suffering for the widows and families followed. Almost overnight, the striking landscape of Thrybergh and Dalton transformed into hills and valleys of tears.

One of the first casualties of the year was that of the Silverwood Colliery manager's only son, Lieutenant Henry ('Harry') Ball. Attached to the Royal Engineers, Harry lost his life on 13 January. Earlier, on 18 March 1915, it was reported that Ball, then serving with the East Yorkshire Regiment, and a qualified mining engineer, had been wounded by a bullet from a German sniper in Flanders. Aged twenty-one, he had only resumed duties with his new regiment for a few days when a mortar shell exploded in or by his trench, also injuring a fellow Second Lieutenant, V.H.M. Barrett, at Givenchy in northern France. Ball was conveyed to the Guards' Cemetery at Windy Corner and buried there.

Detail from the feature in the *Rotherham Advertiser* about Harry Ball. *Rotherham Advertiser*

Dedicated to Lieutenant Henry Ball, this plaque was placed in the north wall of the nave of St Peter's Church by his father, William Henry Ball on 18 June 1916. Brian Elliott

Just five months after Henry's death a memorial plaque was unveiled in the south aisle of the nave, in a solemn ceremony led by Reverend Everson. The congregation sang the Lieutenant's favourite hymn: *How bright these glorious spirits shine*.

Private Enoch Watts. *Rotherham Advertiser*

Another young man, also with a close association to St Peter's, Private Enoch Watts, of the Prince of Wales Own (West Yorkshire) Regiment, was reported as killed in action in November.[20] A Silverwood hewer, Watts was a chorister at St Peter's. The 'no pain' letter from his commanding officer to his widow, Florence, was similar to countless others sent to the next of kin of local miners: 'I am very sorry to inform you that your husband was killed in action on November 1st. He was killed instantly by a shell. I am glad to say he suffered no pain. He had been battalion orderly for about four months and had always performed his duties extremely well. He was well liked . . . Please accept our deepest sympathy in your great loss.' There was no body to bury, Watts one of the 72,246 'missing' servicemen who died on the Somme battlefield and commemorated on the Thiepval Memorial. Newly married, aged twenty-four, Enoch left an infant son.

News about 'Silverwood men' who had 'given their lives for their country' became so frequent in 1916 that at times only portraits of them appeared in the local press. Photographs of Corporal G.(George) F.(Frederick) Willert and privates Ernest Gee, E. (Eli) Charnock, W.(Walter) Woolley and Horace Green appeared in the *Rotherham Advertiser* on 7 October. Although Willert was not listed on the village war memorial, he was a Thrybergh man, born there in 1891, and lived in the village when he was a twenty-year-old pit labourer in 1911. He served with the 6th Connaught Rangers and died of his wounds on 10 September 1916, and interred in France, at St Sever Cemetery, Rouen.

Allen Walker is described as a 'Gallipoli hero', having also served in Egypt before going to France; and also appears to have been a Reservist whilst working at Silverwood. A letter sent to his widow by his hospital nurse mentions a chest wound as the main cause of his death, saying that he 'never grumbled' and always had a 'cheery smile' despite being too short of breath to speak. His widow was also informed that his grave was marked 'with a wooden cross', and she could if she wished 'send out a few bulbs' and he would ensure that they were planted.[21] Walker was Derbyshire born (Chapel-en-le-Frith) and had been working as an ironstone miner at Skelton in North Yorkshire before moving to Eastwood, Rotherham, getting a job at Silverwood. Enlisting in the 6th York and Lancasters on 5 December 1915, he was wounded on 24 September 1916, died in hospital eight days later, and was buried in Puchevillers Military Cemetery in the Somme region of France. His widow, Ada, was left to look after three young children.

By late summer, the *Rotherham Advertiser*, under a sub-heading of 'Another Dalton Soldier Killed', stated that 'Hardly a week passes without one or more of the many brave men who left the Dalton district to take up arms . . . being reported killed'.[22] The 'another soldier' was Private Edwin Fisher, late of Kelvin Street, serving in the Prince of Wales (West Yorkshire) Regiment. The report included an extract from a letter of tribute from his commanding officer describing the circumstances of his death. On 12 August, volunteering to bury the dead bodies of his comrades left stranded in front of his trench, Fisher had crawled towards them when he was shot by a German sniper. He left a widow and three children and was one of three men with the same family name who were killed in action from St Peter's Conventional District in the War.

The *Advertiser* did what it could to maintain morale for the industrial communities in and around Rotherham, publishing cases of gallantry. The family of Edward ('Teddy') Dibbo had moved into Thrybergh parish from Derbyshire via Tinsley (Sheffield). By 1911, parents John and Emily were well known in the area, John working on the pit top at Silverwood as a colliery screen foreman, with three working-age sons, including fifteen-year-old Teddy, employed as labourers. Edward, a leading member of the colliery Ambulance Brigade, and an active student in the Bible Class at St Peter's, had enlisted in the RAMC shortly after the outbreak of the war, still in his

Military Medal holder Edward Dibbo. *Rotherham Advertiser*

Stretcher bearers in the Somme area.. Taylor Library

teens. Attached to the West Riding Field Ambulance, he was deployed as a stretcher-bearer and first-aid man. In a letter to his father, Ernest described how, when carrying an empty stretcher with his pal, the latter was seriously wounded by a shell. What followed next morning however, on 3 July 1916, was an even worse experience, when three shells landed in the dugout that he was occupying, killing three men instantly and burying seven others. Ernest and another mate managed to uncover all seven of the 'badly crushed' men, despite a continuous bombardment of shells, saving their lives. He was subsequently awarded the Military Medal (MM) for his bravery under fire. The names of Edward Dibbo and his older brother, Arthur, are recorded on the Roll of Honour in St Peter's Church.[23]

Dibbo was added to a growing list of former Dalton Main miners who gained gallantry awards in the war, many of them stretcher-bearers. Another, and also MM 'winner', was Private Harry Taylor of the York and Lancasters, who got his award for the rescue of a badly wounded officer whilst under heavy shell and machine-gun fire. The incident occurred at Aveluy Wood right at the start of the Somme Offensive, on the most fateful of days, 1 July 1916, when Harry carried Lieutenant Stanley Else to the field dressing station. Rotherham-born, and said to be a Silverwood miner, Taylor

Harry Taylor's 'splendid deed of heroism' was reported in the *Rotherham Advertiser*, along with his photograph. Brian Elliott

was described as a boarder in the 1911 census, aged eighteen, working as a 'trammer', therefore pushing tubs along the underground workings.[24]

Under the bold heading 'ANOTHER SILVERWOOD HERO', a Whinney Hill soldier-miner from the York and Lancaster Regiment, Private James ('Jimmy') Bullen, 'won' the MM. A former 'rope hand' at Silverwood, and yet another stretcher-bearer, Jimmy, twenty-two, was said to be used to carrying wounded men out of No Man's Land despite great danger to himself. The award was for his bravery in dressing the wounds and carrying his sergeant-major to safety, but was he was badly wounded himself on 22 June, a few days before the 'Big Push'.[25]

Whenever possible, the Rotherham town and parish councils made public presentations of awards gained in the war and several took place in 1916. For 'gallantry in the trenches', Private Edward Gray, of the York and Lancasters, was presented with his MM in October. Private Gray, a Territorial for several years before the war, and a member of the Silverwood Colliery Band, had 'gone out' in 1915, deployed as a stretcher-bearer at the Battle of Ypres.[26]

Towards the end of 1916, news came through of a gallantry award to Corporal Robert Moore of the 8th York and Lancasters, yet another Silverwood miner who got the MM, though he was 'now lying sick in hospital in England ... [in London] recovering from wounds received six months ago'. Dublin-born, residing at Dalton Brook with his wife and five small children, Moore was twenty-six, a hewer, when he had attested at Pontefract in September 1914. His 'wounds' were, like so many others, psychological as well as physical as he was discharged due to 'shell shock by active service' in June 1917. Unable or not wanting to return to mining, Robert opted for employment in farming according to his army records, though he had been a miner for the Dalton Main Company for eight years prior to enlistment. He died aged just thirty-eight, never 'getting over' the horrors of war service.[27]

Corporal Robert Moore, MM. *Rotherham Advertiser*

The most famous and most celebrated example of gallantry by a 'Silverwood man' took place near Meaulte on 3-4 June 1916 when George William 'Bill' Chafer of the 1st East Yorkshire Regiment whose his extreme actions – despite being badly injured (his wounded leg having to be amputated) – resulted in the presentation of a Victoria Cross at Buckingham Palace, on 4 November of the same year. Prior to enlistment, Chafer had worked in the wages office at the colliery.

Mine accidents at Silverwood continued to be reported in 1916. Harry Hall, from Dalton, aged fifty-seven, 'fell to his death' in one of the shafts in January.[28] Another miner, Herbert Jowett, thirty-four, hit by a fall of coal when filling a tub in June, was hospitalised but died nine months later.

1917 : More deaths, more gallantry

Although there were a similar number of known Dalton Main war casualties reported in 1917, for the communities that served the pits the year was as bad as ever. This was due to the much-delayed news concerning the deaths of missing miner-soldiers, particularly from 1 July and the Somme Offensive. Readers had to wait until 30 June 1917 to learn, via the *Advertiser*, that Whinney Hill resident Sergeant Frederick Kelly had been killed on the morning of 1 July, part of a battalion of 'coalminers from the Rotherham district' (8th York and Lancasters), all 'Kitchener's servicemen, trained to the hour and as "hard as nails"'. What a burden of care and financial difficultly it must have been for his widow and five small children.

Several soldiers referred to as 'Silverwood heroes' in 1916 were reported as dead or missing a year later. Military Medal recipient Harry Taylor died on 3 May from wounds received in action on 27 April.[29] Another Whinney Hill MM, Leading Seaman John ('Jackie') Ball, twenty-two, a St Peter's Bible Class member, lost his life on 23 April.[30] Sam Duxbury, a Territorial who had joined up in 1914 and reported missing on 1 July, was now 'presumed dead'.[31] It took *two years* for Enoch Millnes, reported missing on 7 December 1915, to be 'presumed killed' in 1917.[32] What heartache for his family to endure. Like Sam, he had enlisted in August 1914. Thus, the toll of ex Silverwood men killed in action continued to grow during the year. Among those confirmed in the press as dead were Sapper David Thornhill (aged thirty, gas poisoning), Fred Heppinstall (nineteen, pony driver), William Binney (thirty-one, killed in action) and Drummer Herbert Elsey (twenty-one, from wounds).

News of bravery awards were reported on a regular basis. Gunner W. Hodges was awarded the MM 'for conspicuous gallantry' and Corporal Ernest Greaves, 'gassed a year ago' got the same accolade.[33] Other MMs included machine-gunner George Holmes (aged twenty-one), a former pony driver[34] and John Ellis, yet another York and Lancaster soldier who fought on the Somme with the 8th battalion. A stretcher-bearer, Ellis, twice wounded, was said to have helped to convey many injured colleagues to the dressing station whilst under heavy fire.[35]

The Battle of Poelcappelle (Third Battle of Ypres) on 9 October was especially devastating for Rotherham area families because of the number of men wounded, killed and missing *on a single day*, including at least nine fatalities from Dalton Main. Exceptional rainfall had made ground conditions in Belgium terrible to endure for all combatants.

William Binney and Fred Heppinstall were two of the Poelcappelle victims from Silverwood. Another was Thomas Murtagh, aged twenty-four, his family originally from Staffordshire. A machine-gunner with the 4th York and Lancasters, Thomas died after being bayoneted in an attempt to take the high ground of Passchendaele ridge. His illustrious brother, Bernard Murtagh (1/5 York and Lancasters), a former boy miner at Silverwood (according to his own oral testimony), had become a Company Sergeant Major by the age of twenty-one in 1917, having got the MM the previous year; and then Bernard became one of the most decorated of all of the local soldier-miners, awarded the DCM in March 1918, for extreme gallantry and leadership when he took command of his company following the death of its officer. Bernard lived to the grand age of ninety-five, his transcribed reminiscences lodged in Rotherham Local Studies and Archives. He describes his great sadness at losing 'plenty of friends in the battalion', and particularly the loss of his brother Thomas. Not surprisingly, Passchendaele, the scene of his sibling's death, was, by far, his 'worst memory'. It was also where he had a lucky escape, when, having 'sunk up to the chin' in the mud of a deep shell hole, he was rescued from a certain and horrible death by an Australian soldier who dragged him to safety via the butt of a rifle.[36]

Thanks to extant records and news reports, the tragic stories of particular families and their circumstances at this time help us appreciate how they suffered and had to cope with tragic circumstances. Bernard Featherstone, of the 10th York and Lancasters, was killed in action on 28 April 1917. His father George Featherstone had migrated to Rotherham in about 1868, from Staveley, north Derbyshire. Bernard worked as a pony driver at Silverwood prior to enlistment in September 1914, aged twenty. Standing just over 5 feet 5 inches and weighing less than 10 stone, he suffered badly from the cold, waterlogged and crowded conditions of the trenches on the Western Front, 'hospitalised' on five occasions during 1915 (for 'trench feet' and 'chilled feet'), and by May 1916 had gone down with a potentially lethal strain of influenza. By the third week of April he was declared as missing and by mid-May a 'killed in action' note (for 28 April) was added to his records. The York and Lancaster regimental history confirms that Bernard was one of more than fifty NCOs and men killed (or died of wounds) at the Battle of Arleux following a thwarted attack that left very few survivors. He was buried at Chili Trench Cemetery, Gavrelle (Dep du Pas-de-Calais). It was a hugely tragic

The three Featherstone brothers, L-R: Alfred, Bernard and Cyril. *Rotherham Advertiser*

Andrew Featherstone displays Memorial Tablets ('death pennies') for his great-great uncles Bernard and Alfred Featherstone, killed in action in 1917.
Brian Elliott

few years for the Featherstones who had a large family. One of Bernard's younger brothers, Alfred Donovan Featherstone, was also killed in action only a few weeks later, and between 1894-1911 four of George and Martha's children had died in infancy. Perhaps their only consolation was the survival of their now oldest son, Cyril Featherstone, a Company Sergeant Major in the KOYLI's who was awarded the DCM and survived the war.[37]

At home, the miners themselves and the Dalton Main Colliery Company continued to do what they could to ameliorate an increasingly grim situation for so many households. In midwinter, January 1917, the amount of money in support of dependants, subscribed by officials and workmen alone had reached an astonishing £11,755 (about £670,000 today). Apart from regular weekly payments, sums of £10 and £5 were also given to widows and partial dependants in the case of a fatality. The colliery owners continued to allow free rent and coal to the married women of soldiers living in its company-owned houses.

In July, under the headline 'Silverwood Miners Denounce the Pacifists', the *Advertiser* reported a 'mass meeting' of the Dalton Branch of the miners' union in which 'strong indignation' was expressed concerning the Yorkshire Miners' Council recent vote in favour of the Government to begin peace negotiations. Cries of 'Traitors' and 'Pro-Germans' emanated from elements of the assembly, 'men rising to their feet' in order 'to condemn the pacifists'. Cox, the Silverwood delegate, went so far as to denounce the 'so-called miners' leaders' who had advocated a peace, referring to them and their supporters as 'traitors who were worse than the Huns'.[38]

1918-23 : *Armistice and beyond*
The last year of the war saw no marked reduction in the number of reported fatalities of former Silverwood-Roundwood men in combat and associated services, not surprising given the German offensive and Allied counter offensive that was taking place. April was a particularly bad month, and news of the fate of ex-miners appeared in the local press on most weeks throughout 1918. As in 1917, stories about those wounded, missing and killed in action were intermingled with accounts of others who got gallantry awards; and some of the saddest reports were of those who were now 'confirmed dead'

after many months of uncertainty. An August 1914 recruit, Sergeant W. Walker of the York and Lancasters, declared missing on 8 June 1917, was now 'officially reported killed' on that date, another sad example of late news.[39]

Information about the sick and injured, abroad and at home, were also reported on a frequent basis. Among the many in this category were Sapper William Holbrook (Royal Engineers) who was gassed on 1 July 1916 and now in hospital in France,[40] and Private Frank Davis (Royal Munsters), a veteran of Gallipoli and Egypt, who was undergoing hospital treatment in Sheffield.[41]

Reports also uncovered Silverwood miners who were underage on enlistment. Signaller Edwin Clarke, declared missing, had been in the army since the age of sixteen, perhaps influenced by two of his brothers, one of whom was killed in 1917 and the other twice wounded.[42] Private John Marshall (West Yorkshires), killed in action on 20 July, aged nineteen, had served since 1916 'having represented himself older than he was'.[43] Private William Beddows, who had also joined up at sixteen, was sent back to Silverwood pit after parental intervention, but then rejoined the KOYLI's 'on his eighteenth birthday'. After only seven months in France, William was fatally wounded.[44] Numerous other former 'boy soldiers' from Silverwood were reported as killed in action or died of wounds during 1918, among them Harry Dent, George Dowd, John Finney, Ben Hickman, John Newsome, Richard Pritchard and William Sayers.

Among the many war dead are numerous examples that catch the eye when reading news reports, their demise confirmed in more detail after further research. Sergeant Major Donald Alexander Westlake, twenty-nine, died of wounds received in action on 13 April. But he was already a veteran miner-soldier. Westlake had had militia experience as early as 1906, prior to becoming a 'regular', transferring to the Reserve from the York and Lancasters in 1912. Called to service for the Expeditionary Force, the former Silverwood miner fought at Mons and subsequent early battles of the war but was 'shot in the lungs' in 1915. His wounds three years later were even more horrible, shot through both legs and his right arm 'blown away'. Westlake was given a military funeral prior to interment in Masborough cemetery.[45] On the same day that Westlake was fatally wounded, 13 April, Private Joe Cox was killed, though news did not reach his family until August 1919, an extraordinary waiting period for them to endure. Steeped in mining and local politics, Joe's father, Councillor William Cox, was the miners' trade union representative (a 'delegate') at Silverwood.

Another sad example concerns Arthur Orchard, aged twenty-eight when he died, on 29 November 1918, eighteen days *after* the Armistice was signed. From Whinney Hill, Arthur was one of the few survivors of the 8th York and Lancasters' ill-fated Somme experience of 1 July. His subsequent wounding occurred in April, in Salonica, when a Bulgarian prisoner struck him on the head with the butt of a rifle. Arthur died at Radcliffe-on-Trent,

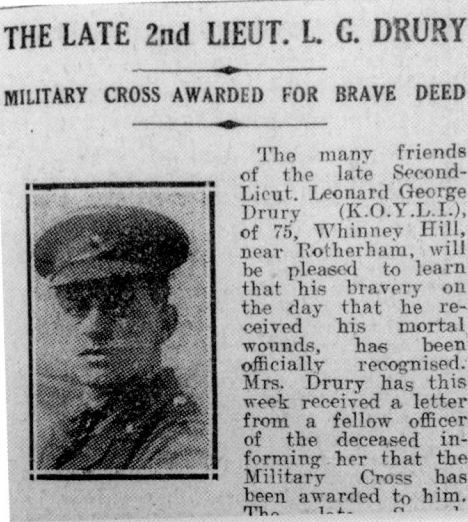

THE LATE 2nd LIEUT. L. G. DRURY

MILITARY CROSS AWARDED FOR BRAVE DEED

The many friends of the late Second-Lieut. Leonard George Drury (K.O.Y.L.I.), of 75, Whinney Hill, near Rotherham, will be pleased to learn that his bravery on the day that he received his mortal wounds, has been officially recognised. Mrs. Drury has this week received a letter from a fellow officer of the deceased informing her that the Military Cross has been awarded to him.

Lieutenant L.J. Drury's MM as reported in the *Rotherham Advertiser*, 17 August 1918.
Rotherham Advertiser

probably in hospital. He would have had a Welsh accent, the Orchards migrating to Whinney Hill from south Wales, Arthur finding work as a skilled miner, a hewer, by 1911; and indeed his army service also involved the West Yorkshire (Prince of Wales Own) Regiment. Buried in Thrybergh cemetery, he left a widow and three young children.[46] Yet another Whinney Hill man, Lance-Corporal Wilfred Back, of the 2nd/4th York and Lancasters, was killed in action on 4 November, only a week before the signing of the Armistice.[47]

Leonard ('Len') James Drury, a neighbour and exact contemporary of Arthur Orchard, was mortally wounded on 10 September 1918. Drury had initially entered service with the Royal Scots Fusiliers as a private, progressing to sergeant-instructor by the time he saw action on the Somme in July 1916; and then to 'office-class', as a Second Lieutenant, prior to a transfer to the 9th KOYLI's. For 'extraordinary bravery whilst leading his men in an attack on a strongly held enemy position' (his commanding officer's words in a letter to his widow), Len managed to gain the vantage point before being shot. His action was honoured by the award of a Military Cross.[48]

Amongst the many gallantry reports that appeared in 1918 concerning 'ex-Silverwood' miners is one about J.[Joseph] T.[homas] Lambley who got the MM in October of the same year.[49] The Lambley family had moved to the Eastwood area of Rotherham via Greasborough, Barnsley and Lincolnshire, J.T.'s father Thomas Ellis Lambley finding employment as a colliery banksman, a very responsible pit-top job, overseeing the conveyance of men and materials up and down shafts. Joseph had moved out of the family home by 1911, aged twenty, living as a lodger in the nearby household of George Wright. 'Joe' was then employed as a miner. A year later, he was married and by May 1915, aged twenty-three, was a gunner in the Royal Field Artillery, having enlisted in the 164th (Rotherham) Brigade. He 'rose through the ranks', promoted to Corporal by the end of his first year of service and then Sergeant in December 1916. Placed on the Reserve after the war, Joe may have returned to work at Silverwood, and certainly stayed in the Rotherham area, until his death in 1956.

Details regarding the bravery of another Whinney Hill miner-soldier, Sergeant Ernest Rollett, were summarised in the local press on 4 July 1918.[50] Held up by German fire, Rollett went alone and killed three of the enemy '[who

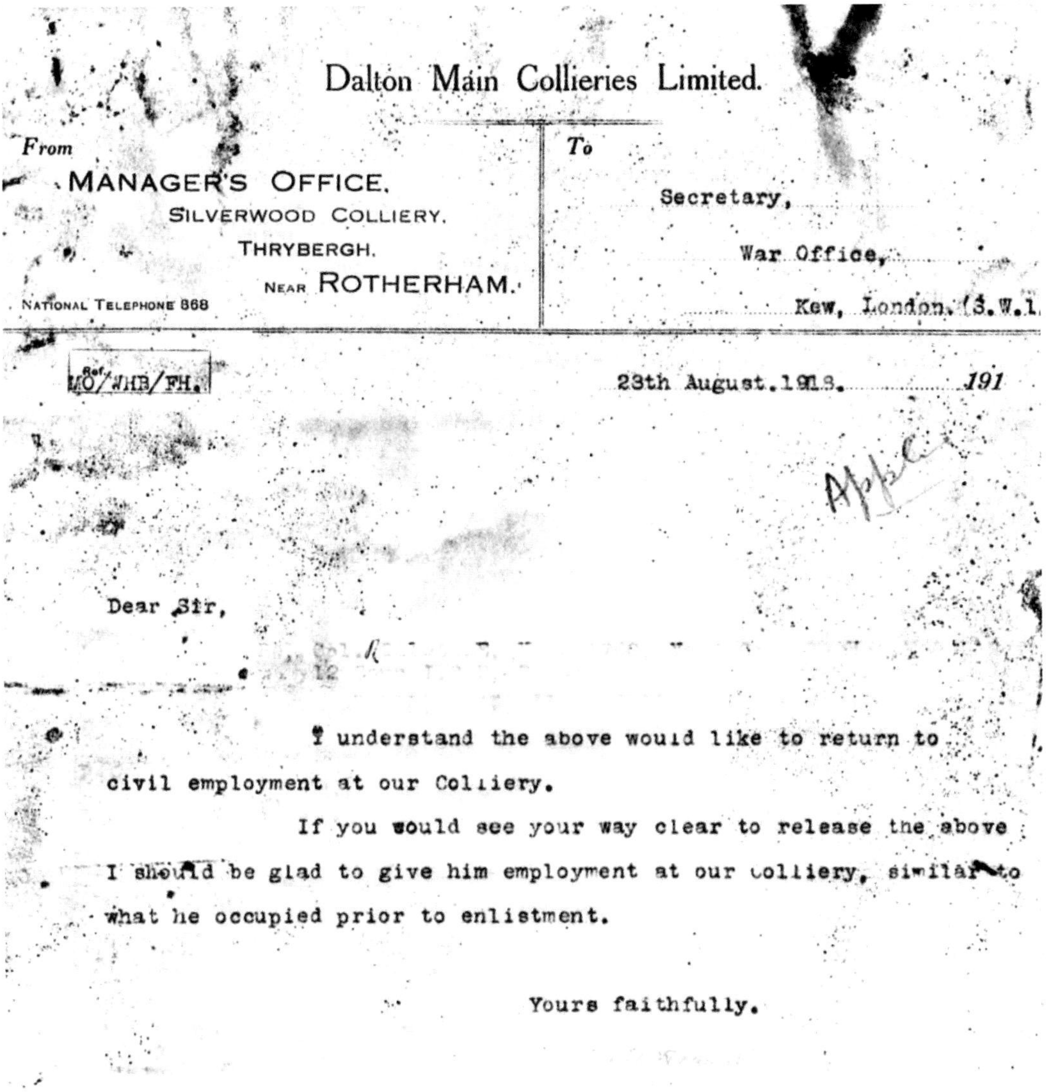

Dalton Main Collieries Limited.

From
MANAGER'S OFFICE,
SILVERWOOD COLLIERY,
THRYBERGH,
NEAR ROTHERHAM.

NATIONAL TELEPHONE 868

To
Secretary,
War Office,
Kew, London, (S.W.1

MO/JHB/FH.

23th August. 1918.   191

Dear Sir,

I understand the above would like to return to civil employment at our Colliery.

If you would see your way clear to release the above I should be glad to give him employment at our colliery, similar to what he occupied prior to enlistment.

Yours faithfully,

Dalton Main Colliery manager's letter to Ernest Rollett. Brian Elliott

were] firing a machine-gun', and seven 'other Germans' then surrendered. Rollett had enlisted, aged twenty, with the 2/5 York and Lancasters on 2 September 1914, transferring to the Royal Munsters until June 1916, then rejoining his old regiment. Wounded in several actions, Rollett may also have returned to work as a miner at Silverwood, the colliery manager's letter, surviving amongst Rollets' numerous military papers.

'War-honour presentations' had become a regular occurrence in Rotherham. In April 1918, Private Samuel Miree of the York and Lancasters was given his MM by the Mayor. The Dalton Lane soldier, despite being wounded in the leg on the Somme on 1 July 1916, and maimed again on 12

August (in both legs and his left wrist), returned to the battlefield. His subsequent service included the carrying to safety of a wounded Lance-Corporal, the action taking place under heavy fire. It was this last piece of bravado that got him the award. Suffering from the after effects of gas poisoning, Miree had been dispatched back to Blighty on 30 September 1917.[51]

Well after the Armistice, prestigious gallantry awards continued to be presented to former Silverwood miners, with as much officialdom as befitted each occasion. Private Alfred Scholes (8/York and Lancasters) received his DCM (and Russian Order of St George) in the spring of 1919.[52] 'Under shell fire', Alfred was reported to have carried to safety a wounded comrade at Orvillers, on 1 July 1916. His Russian award was given via the Red Cross 'for bravery as a stretcher-bearer'. The delay in the presentation was understandable due to Scholes's long service, which included his capture by the Germans and incarceration as a PoW from April 1917.

Another 'foreign honour' went the way of William Marriott, who had enlisted, probably under age, when he was a fitter (mechanic) at Silverwood Colliery, in December 1915. William's abilities were soon realized as he rose to Sergeant status in 3/York and Lancasters. Drafted into the Machine Gun Corps, he saw action in Salonika, most notably when a bomb burst near him, killing three of his crew, leaving himself and a comrade, a Lance-Corporal, shaken but alive. Despite this 'against the odds' situation, the pair managed to open fire and 'execute' many of the German opposition. Marriott, said to be 'only 22' was one of twenty-three ex-Silverwood brass 'bandsmen' who had joined the Colours, almost half of them 'gaining distinction'.[53]

Fatalities in the Silverwood mine continued to be reported in varying detail in the press. Mr A. Blenkinsop of Dalton House, the colliery agent for John Brown & Company, was killed in a roof fall 'whilst inspecting the pit' in 1918.[54] Not long afterwards, an inquest was held by the Rotherham Borough Coroner concerning the death of Robert C. Brocklesby, aged forty-seven, who had died in hospital following an underground accident. A fall of coal had struck Robert on the head whilst he was filling a tub, fracturing his skull. An 'accidental death' verdict was given, quite usual for such events where the exact cause or causes were not fully evident or explained.[55] What the news reports did not reveal was that Brocklesby had a military record stretching back to 1890, and at the age of forty-three in 1915 he had enlisted in the 19/Sherwood Foresters. His later war service was, however, limited to about eighteen months as he was transferred to the Reserve, his employment getting coal at Silverwood Colliery taking precedence. It turned out to be a most ill-fated 'escape to Blighty' for the Lincolnshire-born Brocklesby, getting killed in the pit. He left a widow, Mary and five children.

Earlier, in April 1918, there was a stark reminder that pit accidents continued to impact on the lives of teenage miners and their families. John William Wainwright, a fourteen-year-old pony driver with only four weeks underground experience, lost his life after being crushed by out-of-control tubs, pulled chaotically by a frightened runaway horse. A 'locker' (short brake pole) had not been properly placed in the back wheel by his pony-driving workmate. An 'accidental death' verdict was returned at the inquest.[56] Wainwright came from a poor family background, his father Harry working as a labourer and 'scavenger' in 1911, when there were two other young children, a son and daughter, resident in their small property.

By October 1918, the 'Silverwood Colliery War Relief Fund', accumulated from a voluntary weekly wages levy from officials and workmen had almost doubled in little more than a year, the total now standing at a formidable £20,433 (about £885,000 today). Most of the Fund had already been distributed to soldiers' dependants and families, so it was an efficient process. Also noted in the same report was that 1,276 men had enlisted to date, and that at least 200 employees at the colliery 'had made the supreme sacrifice'.[57]

As at other pits, times were difficult at the Dalton Main collieries immediately after the war, as returning former miners were by no means guaranteed employment. However, the demand for British coal was strong right through to 1921 when the market collapsed. Thus there was a healthy drift of people (even a rush at times) into the Silverwood catchment area after the Armistice, some of the men already experienced miners but many more with other occupations. By the early summer of 1920, reductions began to appear in coal owners' profits, the 'Silverwooders' themselves – about 4,000 of them now – placed on 'exceptionally short time', according to one report.[58] Despite falling demand, and against the odds, the West (No.1) Pit of Silverwood achieved a record output in September 1921, the achievement getting a mention in the national press.[59]

* * *

A survivor

Despite the significant loss of life of those Dalton company miners who joined the forces, the bulk of enlistees returned home to find work back in the mining industry or elsewhere. This does not mean to say of course that they were 'unscathed', as a good number are likely to have had experienced some form of physical or 'invisible' disability that stayed with them for the rest of their lives. Let's take a look at an actual example, that of a young miner, Ernest Jones.[60]

The Jones family, like so many others who found work in south Yorkshire mines, were incomers. Ernest ('Ernie'), though born at Masbrough, near Rotherham in 1895, had a Staffordshire father, listed as a carter (of coal) in the censuses. The second oldest of nine children, by the age of thirteen Ernie is believed to have been employed at Silverwood, about three miles east of his home. He was 'set on' as a pony driver, hauling tubs between the workings

This photograph taken in a Rotherham studio shows Ernest Jones (right), during a home visit, alongside his older brother, 'Frank' (Francis), in full uniform, who served in the Royal Horse Artillery. The two stripes on the wrist area of the sleeve of Ernest's coat suggests that he was wounded twice. Leslie Jones

and the pit bottom of the mine. Training was minimal, not helpful where some ponies were awkward to handle. A variety of accidents might happen. Indeed, Ernie got a serious back injury whilst 'driving', but that did not deter him, aged only nineteen and just 5 foot 3 inches tall, to join the Colours. He enlisted in the 5th York and Lancasters, at Rotherham on 2 January 1915. Two years later he was in France and Flanders, deployed in a very important strategic role, as a signaller at Ypres and Passchendaele. His service record shows that he was wounded, shot in the arms and legs, on 3 May 1917, and dispatched back home a few days later.

Colonel Wylly's regimental history of the York and Lancaster Regiment provides us with insight as to what happened on and around the day of Ernest's wounding, which took place during an attack on the Hindenburg Line when heavy losses occurred:

> At 5.30pm on the 2nd [May] the Battalion (2nd/5th York & Lancs) moved from its camp at Gomiecourt to a rendezvous at St Leger, preparatory to the attack; and as stated in the War Diary, 'the Battalion attacked the Hindenburg Line, the 2nd/4th York and Lancaster on the right; the 2nd/5th York and Lancaster attacking as second and third objectives. . . respectively . . . the attack was not successful, although the greater part of the Battalion got into the Hindenburg Line and many into the Support Line; those who obtained their objectives put up a gallant fight, and were either killed or wounded, a few withdrawing at nightfall'.[61]

Jones's convalescence was brief as he was summoned to active service less than two weeks later, continuing until demobilization in January 1919. Later, Ernie appears to have abandoned mining, working as a general labourer after moving to a suburb of Sheffield with his wife Eleanor, whom he had married

This Roll of Honour 'of Men connected to St Peter's Church' lists 114 names, including twenty who lost their lives in the 1914-18 war. Most lived in the small community of Whinney Hill. Brian Elliott

in 1920. But at the age of almost forty-seven the 'old soldier' volunteered for the Pioneer Corps, serving in north Africa under Montgomery for the last three years of the Second World War. His son, Leslie, told me that his father never talked about his war experiences.

Silverwood: A Wartime Colliery • 105

Chapter Five

1916: UNSPEAKABLE HORROR

Miners were involved in all the main theatres of war during 1916, on land, at sea, and to a far lesser extent, with the Royal Flying Corps. But for most it involved infantry service on the Western Front in France during the almost year-long struggle on the Somme. For trivial gains, the Battle of the Somme lasted forty-two days, and was responsible for the heaviest loss of life in a single day's fighting in British military history: 19,240 dead on 1 July 1916, and 57,470 casualties by November.

In coalfield communities over the ensuing months columns of named casualties and of the missing permeated the war pages of local newspapers. As the lists lengthened and there was no end in sight an air of depression weighed heavily on working miners and their families. The most dramatic evidence of the scale of slaughter occurred when front and subsequent pages contained portrait images of miners and their pals killed in action.

More than a century later the losses on the Somme are difficult for all of us to comprehend, impossible to appreciate fully no matter how many statistics are presented and explained. It's the individual stories, the 'mini-

biogs' of the soldiers and their families, usually well researched by family and local historians, that provide some meaning to what it was like. This chapter contains examples relating to the Somme Offensive (the 'Big Push'), its prelude and latter stages.

It is not known how many ex-miners were classed as casualties during the Big Push but thousands were certainly participants in the various battles and attacks, from Serre, Delville Wood and High Wood to Flers-Courcelette, Morval and Ancre, when winter rain and snow put an end to hostilities.

The miners often served with their mates in Pals battalions, and in numerous 'collier regiments' such as the Durham Light Infantry, York and Lancasters, King's Own Light Infantry (KOILIs 'Miners' (Pioneer) Battalion'), the Monmouthshire Territorials, Swansea Pals (14th (Service) Battalion Welsh Regiment and Northumberland Fusiliers (Tyneside Scottish). Those that had been targeted for the Royal Engineers' tunnelling companies operated in contexts that included some of the bravest and most unselfish actions imaginable.

Many others served in 'logistics' units, in the new Pioneer battalions that supported the Royal Engineers, and also in the developing Tank Corps. Not to be forgotten were the numerous 'short-stature' miners in the English and Scottish Bantams who fought on the Somme. And absolutely vital were the former pit 'ambulance' men' and 'first aiders' recruited into the RAMC or serving as regimental stretcher-bearers, alongside a scattering of 'miner-conchies' who – often unarmed – rushed to help stranded colleagues on No Man's Land or used their medical skills at field hospitals. These 'battlefield medics', took on roles not unlike those that adopted in the aftermath of a large mine disaster. Attending to the dying and the dead must have been the most horrendous and life-affecting experiences of all their duties.

Saturday, 1 July 1916

Great swathes of infantrymen were mowed down on the terrible walk, stumble and crawl across the crater-strewn killing fields of the Somme, from 7.30 on a summer Saturday morning. Of course not all were miners, but many of those

in the units referred to above were killed or badly injured, others suffering for weeks, months and years afterwards. The casualty toll is astonishing. Most of the Durham Light Infantry battalions, many of them 'miner volunteers' incurred many hundreds of casualties, 2,469 officers and men by the end of the year. The 15th DLI were the worst affected, with 171 in July and 345 by December. Of the 720 Accrington Pals who 'went over the top', 584 'failed to appear' by 2 July or required urgent medical help. The Leeds Pals lost 750 of about 900 of its advanced participants, and the Sheffield's City Battalion lost about half of its men. At the first roll call, 197 officers and other ranks of the 12th KOYLI's Miners' (Service) Battalion – normally involved in trench, dugout and related work – were 'unaccounted for', mostly killed and missing. The 8th Battalion KOYLIs combined with the 8th York and Lancasters lost at least half of their men after braving the machine gun fire when crossing No Man's Land; and the 9/York and Lancasters that followed them suffered at least 581 losses from a complement of 761 officers and other ranks. The four Service battalions of the Northumberland Fusiliers (20th-23rd Tyneside Scottish) also suffered badly, each losing between about 560-670 of their 'rank and file' within hours, as well as every commander and more than eight officers. The two Barnsley miner-dominated Pals battalions (13th and 14th York and Lancasters) got battered at Serre, with over 500 casualties. The Swansea Pals had about a hundred men killed during their attack on Mametz Wood and 600 were lost by the end of the Big Push. Actually and symbolically, for most of the volunteers it was the end of a two-year dream. Any idealism remaining at the start of 1916 evaporated like a pool of water in a hot desert and by summer the enthusiasm of those who had volunteered had gone.[1]

Underage survivors remember[2]

Veterans who did – eventually – talk about their Somme experiences had amazing stories to tell. **Frank Lindley**, 2nd Barnsley (14th York and Lancasters), aged sixteen, said this about his precarious, stranded position near Serre:

> There I lay studying what to do next when this 'ere 'whizz bang' came over and that put paid to the lot of us in that shell hole. It came over and burst – I never heard it coming. A piece of shell went through my thigh [and] took my trousers with it. When I looked down my trousers were gone and blood was pouring out and a piece of shell was sticking out of my thigh. I started to roll back through No Man's Land towards our trenches. When I neared John Copse I found a sap and fell into it. There was another bloke there with a wound. I lit a cigarette. Another shell burst close by, as the Germans were still targeting the copse. The other fellow was hit [on his] arm and a piece of hot shrapnel burnt a small hole in my leg. With disaster on the scale it was you wondered how the bloody hell they were going to clear this mess up.

Frank Lindley, was only sixteen years old when he was in action on the Somme at Serre.
Taylor Library

Lindley had enlisted at the age of fourteen, convincing his recruiting officers by his stature and attitude that he was twenty. Although not a miner (he had what he later called a 'posh' job as a dental technician), he joined the Barnsley Pals from his home area of Sheffield, even though he had previously enlisted with the Royal Artillery. His desertion was because he felt that serving alongside tough Barnsley miners would give him a better opportunity to avenge the death of his older brother, Able Seaman Harry, who had lost his life in an action involving HMS *Hawke*, in the North Sea in 1914. When I worked at Wickersley Comprehensive School, near Rotherham, in 1987, Lindley came along to talk about his war experiences to a group of students. He passed away a year later, aged eighty-eight.

Harry Hall, 1st Barnsley's (13th Battalion, York and Lancasters), aged nineteen, was very lucky since he advanced and then retreated relatively unscathed, at least physically:

> We started working our way to the left and right to see how far we could get . . . We dare not go any further . . . If you had put your head up there was another line of trenches with Germans still firing. There wasn't much we could do on our own, because it was only a small

Harry Hall, takes a rest. He got promoted to Corporal despite his youth. Brother Tom, a KOLI, lost his life on the Somme on 1 July and another brother, Arthur, was killed at Gallipoli. Taylor Library

section we had broken in to. We were taking enfilade fire and we had no cover. The word was passed along to go back as quickly as we could. I don't think anybody could have got further, especially into Serre, the Germans were too strong.

Harry was seventeen, a young miner from Houghton Main, when a 'creative' recruiting officer allowed him to enlist in the 1st Barnsley Pals. Two of his brothers, Tom and Arthur, lost their lives in the war, the former with the 9th KOYLIs on the Somme on 1 July, and the latter at Gallipoli.

Tyneside's 'miner-battalions'

The village of Ryton-on-Tyne, situated seven miles west of Newcastle and south of the Scottish borders expanded rapidly from the middle of the nineteenth century after the development of Greenside, Stargate and Emma collieries – under the ownership of the Stella Coal Company. By the start of the Great War well over 2,000 men and boys were employed at the three pits, about 750 of them at Emma. But by 1915 so many miners had answered 'Kitchener's call' that Stella's Ryton workforce was reduced by almost a third.

The great influx of miners from the North made the recruiting office at Newcastle-upon-Tyne the busiest in the country. Here, groups of men from the same colliery queued for several hours prior to passing through the stages of enlistment. Newly-formed Pals battalions filled quickly, well in

Sir Arthur Markham (1866-1916)

Coal owners and miners were not the best of friends during and after the Great War years. However, some did gain a lot of respect from the communities in which they operated because of their social support of their employees. Sir Arthur Markham took matters a step further. When serving as the Liberal MP for the coal-town of Mansfield he was the loudest and the most influential voice against the enlistment of 'under-agers', boys from the mines and elsewhere who were accepted into the Colours. Markham's independent stance in Parliament and on public platforms mirrored the honest and forthright attitude of many of miners that he employed. One reflection of this was the gratitude displayed by the miners' unions in their attendance at his funeral.

Such was Markham's campaigning that it was thought that it was this tireless passion that contributed to his premature death at the age of forty-nine in 1916. Markham's wife was also a key figure, working as a matron when their family home was transformed into a military hospital.

Sir Arthur Markham, coal magnate, served as MP for the Nottinghamshire constituency of Mansfield from 1900 until his death on 5 August 1916. Brian Elliott

From a privileged background, Markham had inherited the family-named Markham Colliery in north Derbyshire from his father. Becoming a formidable industrial entrepreneur in his own right, he was to own and manage coal mines from his base in Derbyshire to south Wales; and in south Yorkshire he was responsible for the sinking and development of what became Britain's biggest pit: Brodsworth, near Doncaster. With his sister Violet, Sir Arthur was instrumental in creating one of the best examples of social provision for mineworkers: Woodlands Model Village. Although the pits have long gone, the Markham legacy remains as testament to their influence via numerous landscape features and street-names.

---

excess of the 1,000 of men required for a battalion. In October 1914 the new Tyneside Scottish (20th Service) Battalion of the Northumberland Fusiliers reached its complement within ten days, so three more cohorts (22nd-23rd) were commissioned (to Brigade strength) – and completed within a month.

Most of the new North East battalions saw action on the Somme battlefield and suffered terrible losses. The Tyneside Scottish and Tyneside Irish lost half or more of their number on 1 July 1916. For these, an indication of the losses incurred by their four Pals battalions can be seen by reference to the analysis in Stewart and Sheen's *Tyneside Scottish* book, Sheen's *Tyneside Irish* and, online, via the Lochnagar (Crater) memorial site.[3]

Emma (above) and Greenside collieries (below) employed a combined workforce of about a thousand men and boys in 1915, part of the Stella Coal Company's considerable mining interests in the Ryton district. Michael Hardy

Overall, locally, an indication of the carnage on the Somme can be seen by the presence of so many 'pit village' war memorials in County Durham, including several coal-owner and workmen-sponsored examples. Ryton's war memorial – in one of the smaller villages – has 273 known names inscribed for the Great War.[4] It's impossible not to notice several surnames listed multiple times, clusters of bereavement hard to imagine for the Armstrong, Charlton,

Graham, Johnson and Richardson families especially. Unveiling ceremonies also took places in nearby pit villages, churches and chapels, adding to the weight of sorrow in the Ryton district. The workmen at Emma Colliery raised funds in order to erect a Memorial Hall in 1923-24, and a commemorative bronze plaque was placed on the frontage recording 58 of its 'workmen and workmen's sons' who lost their lives in 1914-18.[5]

All the heartbreak is nigh impossible to comprehend unless personal profiles are uncovered. Below are just two examples, extant courtesy of Ryton and District War Memorials Project, with details supplemented by my own research.

On Monday 20 November 1914, **Arthur Tremble**, a 34-year-old hewer from Emma Colliery, with at least twenty years of underground experience, left his home in Ryton and made the seven-mile trip to Newcastle's recruiting office. Like many of his mining mates he had 'swapped his pick for a rifle' and joined the 'Tyneside Scots', in particular the 22nd (Service) Battalion (3rd Tyneside Scottish) of the Northumberland Fusiliers. Arthur may have felt that his decision was made so as to escape from the dangers and the discomforts – as well as the unpredictability of his wages – for a more settled, more exciting interlude; but like many others he was also part of a wave of genuine patriotism that also emanated from the local pits. Being among mining mates from similar backgrounds was also a persuasive factor, certainly one that eased any anxiety. His wife Clara, however, probably worried a great deal about his decision, as she was left to look after two small children on her own.

Private Arthur Tremble.
Royton and District War Memorials Project

In France from January 1916, Arthur gradually advanced towards the Somme, reaching the British front line to the east of the Bapaume-Albert Road and close to the strategically important but tiny French village of La Boiselle, defended by a well German entrenchment overlooking British positions. Arthur's 22nd Battalion were to advance behind the 21st (with the 26th Irish behind them and the Lincolns, Suffolks and 24th Tyneside Irish to the east, their immediate right), at 7.30 am on 1 July, two minutes after the most spectacular detonations of the Great War, most notably the great mine explosion that created the vast Lochnagar Crater (see p.116) which took the Germans by complete surprise. At the same time the 'Y-Sap' mine (known by the Germans) just west of La Boiselle, by the road, was detonated. Arthur witnessed both blasts, unbelievable in their scale and volume of sound, great masses of debris falling from the sky like a storm of volcanic bombs. Then, he was part of the over-the-top trudge, stumble and crawl northwards, amid a

German front line trench in the hollow of Mash Valley between Ovillers and La Boisselle. Taylor Library

blaze of German firepower. It is not known how far he managed to reach, but he lost his life in No Man's Land – when advancing or during the retreat. Like so many of others, his body was never recovered from the killing fields of La Boisselle.

It may have taken weeks or even months before official confirmation of his death reached Clara back at Ryton. Arthur Tremble is remembered at Theipval Memorial for the Missing of the Somme and locally in Ryton where many widowed families continued to be dependent on the goodwill of the Stella Coal Company.

Arthur Tremble would have known a fellow miner-enlistee from Emma Colliery, **Richard Hogg**. 'Dick' as he may have been more popularly known, was a little older than Arthur, thirty-six in 1914, and had far more responsibility as he was already described as a 'Deputy Overman' in the 1911 census, a senior official at Emma, in charge of regulating underground operations and the oversight of teams of colliers. He was also a trained St John's Ambulance man, so used to dealing with emergency first aid, rescue equipment and safety procedures.

Born into a large Gateshead area mining family, in Browney Colliery village, Hogg's life was ingrained in pit work from at least the age of thirteen or fourteen. Married to Catherine in 1898, he left mining for a short period almost three years afterwards, enlisting at Newcastle into the Imperial Yeomanry – via a short-service attestation – on 10 January 1902. Standing only 5 feet 4 inches and weighing less than 10 stone, his battalion was part of a third contingent of volunteers that saw limited action towards the end of the Second Boer War.

Hogg opted for discharge a few days after arriving home, on 28 October 1902. Desperate to earn 'pit-money' to support his wife and growing family, he moved from Browney to Ryton, settling at Crawcrook, close to his new workplace, Emma Colliery. After twelve years, in November 1914, now father of ten children, Richard presented himself at the Newcastle recruitment office for a second time. He was welcomed into the new 20th Battalion (1st Tyneside Scottish) of the Northumberland Fusiliers, and allocated the service number 20/248.

Given his mining and medical background it was not surprising that Richard Hogg was used as a 'battlefield medic', a stretcher-bearer, in the wake of his pals as they advanced across No Man's Land – the infamous 'Marsh Valley' – in the direction the German front line, just west of La Boiselle and close to the main road. Like Tremble, he would have witnessed the great detonations just prior to 7.30am and was no doubt soon required to assist injured and dying comrades. It would have been in this guise, probably with a mate and having little or no protection, that he lost his own life.

What a massive blow it must have been for Catherine and the Hogg family when news eventually reached her that Richard had been killed in France, on 1 July 1916, aged thirty-eight. Like Arthur Tremble and so many

Lochnagar Crater

Today, walking through parts of the pleasant farmed landscape of the Somme it's hard to imagine what a devastated topography it was in 1916-17. At the edges of fields, it's quite easy to find pieces of shrapnel lying on surfaces. Goodness knows how many millions or, more realistically, billions of military fragments remain on and below farmland on one of the most battered and undermined landscapes anywhere on earth. It is, however, above all a 'respected landscape', a huge unmarked cemetery containing the remains of so many soldiers of the Great War.

Fortunately, small areas such as the preserved and protected battlefield landscape at the edge of La Boisselle village – and the associated archaeological and research of the site – provide us with a better appreciation of our First World War ancestors, particularly the professional miners who were at the apex of operations. And, close by, is the unmistakable presence of the Lochnagar Crater.

Anyone visiting the Lochnagar site cannot fail to be impressed by the sheer size and grandeur of the place, as if a massive meteor had crashed there. The huge cavity – 300 feet wide and 70 feet deep – was the climax of the work of teams of miner-tunnellers, created on 1 July, the first day of the Somme Offensive in 1916.

The Lochnagar memorial and site attracts thousands of visitors of all age every year.
Brian Elliott

The tunnel that preceded the blast was started seven months earlier by the 185 Tunnelling Company of the Royal Engineers and completed in March 1916 by the 179TC.

Thanks to the great foresight of its owner Richard Dunning, Lochnagar is preserved and protected, a place that has been seen by many thousands of visitors, including those with direct genealogical links to the former miners and infantry men who survived or lost their lives in or near this area. Commemorative plaques, sponsored by individuals and families, bear testimony to their remembrance as does the annual Remembrance Ceremony on 1 July, organised by the Lochnagar Crater Foundation.[6]

others from the North East having no known grave, Private Richard Hogg is commemorated on the Theipval Memorial. His allocated Victory and War medals – and inscribed 'Dead Man's Penny' (Memorial Plaque) – seem so trivial for such a huge sacrifice. Worse, however, was to come for his grieving family.

Catherine Hogg, late wife of 'war hero' Richard – with one of the largest families in the Ryton area – was threatened with eviction from her home, owned by the Stella Coal Company. It was only by taking in lodgers that she was able to scrape enough cash together to pay the rent. Goodness knows how she coped in such crowded circumstances. No doubt neighbours helped, as they always did in coalfield communities. But there was no respite in the Hogg family's grief as Richard's youngest brother, William Armstrong Hogg, aged twenty-six, a former Emma hewer, was killed in action on 9 August 1917, his remains also never found.

On the most fateful summer morn, 1 July 1916, **James Gardiner** (aka

Detail from a studio photograph of Private James Gardiner, attached originally to the 4th Northumberlands and then serving in the 24th (Service) Battalion of the same regiment (Tyneside Irish). Paul Atterbury / Philip Gardiner

Gardiner's memorial at Lochnagar Crater still shows his misspelt name, used on surviving military records. Lochnagarcrater.org

Gardener)[7] and his 1st Tyneside Irish's pals had waited a few anxious minutes entrenched in a second line prior to their advance across No Man's Land towards La Boisselle. Facing Lochnagar, under instructions, the commander had delayed their assault until after the great hail of debris from the anticipated great bang had settled. Problem was, this gave the German machine gunners more time to confirm their own positions and fire a hail of bullets towards their attackers. The walking cohort stood little chance of survival. It's likely that James lasted but a few moments in the maelstrom. Indeed, within ten minutes most of his mates were also dead or dying. His 24th (Service) Battalion of the Northumberland Fusiliers suffered 634 casualties that morning including 147 killed, a terrible toll. 'Gardener' became one of 513 Tyneside Irish names inscribed on the Memorial to the Missing of the Somme at Thiepval. James, like many of his mining mates, may have attached his identity tag on his braces rather than around his neck, a tradition adopted by many miners who were used to doing the same for their pit 'checks'. But, if bodies were found the absence of a neck tag may have meant classification as an 'unknown', and a more careful search not made. Sadly, about 85% of missing soldiers were placed in this category.

From an Irish background, James was born at Tanfield, County Durham in 1886, and married Ann Moat in 1910. He resided with Ann and his in-laws at the time of the 1911 census when he was described as a 'hewer and putter', so would have got the coal and helped its haulage. Unfortunately their first two

This photograph was taken just one minute after zero on the 1st July 1916 and shows a support company, belonging to one of the Tyneside Irish battalions, moving forward towards the highest ground on Tara Hill. Within seconds of this scene being recorded many of these soldiers became casualties as they cleared the ridge and were exposed to the machine guns firing from positions within and to the rear of La Boisselle village.

children died in infancy and Ann was left with a baby that her late husband never saw. Private James Gardiner died on the Western Front at the age of thirty.

As mentioned above, the combined casualties sustained by the Northumberland Fusilier battalions during 1-3 July 1916 were among the highest for any brigades involved in the early days of the Somme battle. The facts that emerged were brutal. Only a hundred men from the 23rd Tyneside Scottish, for example, answered their names at roll call.[8]

The war underground

Mining engineers

Mining engineers usually came from middle-class, better-off backgrounds compared to 'ordinary' miners, but as part of their professional (often university-based) studies had to experience what it was like to work below ground, liaising with officials and the miners themselves. 'Serving their time' also involved getting to know – at first hand – the technicalities of mining and the application of mine safety and mine rescue. Practical experience of the latter made them ideal candidates for employment as mines inspectors. Henry Moore Hudspeth was thus a 'perfect target' for a commission in the Royal Engineers. From 1916, he was a prominent figure in tunnelling strategy and operations on the Western Front.

Norton Griffiths had recommended to GHQ that mining engineers should be placed in command of the new tunnelling companies and this was widely enacted from the autumn of 1915. Such men, specially commissioned into coalfield area regiments such as the Monmouthshires – unlike the tunnellers themselves – were often rewarded with gallantry awards. William Clay Hepburn, manager of Oakdale Colliery in the Sirhowy Valley of south Wales, was the first civilian mining engineer to command a tunnelling company of the Royal Engineers, the 172TC, when he was attached from the Monmouthshire Regiment. Captain (later Major) Hepburn got the DSO (Distinguished Service Order) for his achievements. Arthur Edwards, son of the general manager at Bleanavon Iron & Coal Company, a Captain in the Second Monmouthshires, and part of the British Expeditionary Force from November 1914, commanded A Company during the Somme battles. Edwards was awarded the MC (Military Cross). Captain Cecil Cropper, a 'British' mining engineer (actually born in New South Wales, Australia) serving in the Northumberland Fusiliers, was attached to the 173TC during 1915. He also got the MC, and after taking command of 250TC, was awarded a DSO in 1918.

The 171 Tunnelling Company (171TC) 1915-1917

One of the earliest tunnelling companies, miners from the coalfield communities of south Wales, as well as those already serving in the South Wales Borderers, Welsh and Monmouthshire regiments were prominent

Major H.M. Hudspeth (1886-1971)

Major H.M. Hudspeth.

From Willington, County Durham, Henry Moore Hudspeth was the son of a colliery clerk and cashier employed at the local colliery, owned by the Strakers and Love partnership.[9]

It was at Willington and Brancepeth collieries that he worked as a trainee mechanical and mining engineer, in 1908 graduating with a BSc at the University of Newcastle's famous 'colliery school', Armstrong College. Taking advantage of a Daglish Travelling Scholarship, Hudspeth extended his mining experience for a year, in Europe. Only in his early twenties, he was elected as an Associate member of the North of England Institute of Mechanical and Mining Engineers, and in 1911 was appointed as a HM Junior Inspector of Mines for the Yorkshire and North Midland Division. Hudspeth narrowly escaped death in 1912, during the ill-fated rescue operations at Cadeby Colliery, near Doncaster, following an unexpected further explosion that cost the lives of fifty-three volunteers.[10] After working in Leeds and Doncaster, in 1915 he was called to serve with the Royal Engineers and was soon supervising and directing tunnelling operations with the 171TC in France.

Today, Spranboekmolen, West-Vlaanderen, Belgium (aka Lone Tree Crater) is a picturesque landscape feature, dubbed the 'Pool of Peace' memorial site, a popular venue for visitors to the Ypres Salient battlefields. It was here, belatedly, on 7 June 1917, that 90,000 pounds of ammonal was detonated, created and placed by the 171TC of the Royal Engineers a year earlier. It is owned and cared for by the Toc H (Christian) organisation, at Poperinghe.
Brian Elliott

> Whilst an acting commander, the knowledge, actions and demeanour of this 'stocky, energetic north-countryman' was such that in March 1916 Norton-Griffiths composed a letter to his mother, extolling her son's 'invaluable work' and assuring her that he was 'in the best of health, jolly and looking very fit'.[11]
>
> Promoted to the rank of Major and given total command, Hudspeth's duties now included the heavy responsibility of 'signing off' the final chamber at the end of the long gallery under the Messines Ridge, 88 feet below the Germans' powerful position at Spranboekmolen (Lone Tree Crater). This was completed after seven months hard and precarious work, on 28 June 1916, though not detonated until almost a year later, on 7 June 1917. Whether or not Norton-Griffiths's 'jolly' remark was an accurate appellation for Hudspeth, he did enjoy celebrating the tunnel's completion, with toasts of champagne with a fellow officer in the specially created 'explosion chamber' at the gallery end.[12] The achievement of Hudspeth's 171TC was of considerable importance in the context of the Messines offensive and its aftermath.
>
> Leaving the army with a DSO and MC – and a master's degree in recognition of his previous academic success – Hudspeth was targeted for senior posts within the Mines Inspectorate and served as Chief Mining Engineer to the Safety in Mines Research Board.

recruits.[13] It was in the Hill 60 area of the Ypres Salient that the 171TC (and 172TC) engaged the Germans in the first major British offensive mining operation on the Western Front, despite many of the 'soldier-tunnellers' having minimum military training. By July 1915, the Company had moved to Ploegsteert and led the long, drawn-out and geologically difficult underground offensive that was a vital part of the undermining of the Messines Ridge during 1916-17. Earlier, in April 1916, the 171TC had taken over from the Canadian tunnellers at the most important strategic site occupied by the Germans: Spranboekmolen. Here, the existing tunnel was driven to 1,717 feet after seven months of subterranean work. A huge charge of 91,000lbs of ammonal was completed by the end of June 1916. Delays were due to German counter-mining of 'subsidiary' tunnelling efforts, and the highly dangerous working conditions, hampered due to frequent 'seepages' and 'blows' of gas. Remarkably, with only a few hours to spare, the miner-tunnellers of 171TC were able to overcome all hazards, enabling the detonation to proceed – as planned – for the Messines 'big bang', at 3.10 am on 7 June 1917. This contribution of the 171TC is summarised by the historian of the south Wales 'miners at war' as 'a testimony to their skill, bravery and sheer "bloody-mindedness" in getting the job done'.[14]

Bedson's extraordinary escape from the dark
In the prelude to the start of the Somme Offensive there were many instances of miners involved in extreme underground activities. Ordinary men finding themselves in extraordinary circumstances. Here is just one quite amazing

story concerning William Henry Bedson of the 250TC. He was given up as 'long dead' when, in mid-June 1916 – on a 'bodies recovery' mission – he was found alive, well and 'with his humour intact'.[15]

A deep tunnel through the clay had been started at Petit Bois in the Ypres Salient in late 1915, the prelude to the Messines Ridge offensive. The underground workings were detected by the Germans who set off two heavy charges on the morning of 10 June, entombing the twelve Royal Engineers, all former miners. Organized by Lieutenant William Hayden Reece, a Welsh mine surveyor experienced in colliery rescues, shifts of experienced miners wearing Proto breathing gear commenced explorations in seemingly hopeless conditions. After days of toil, extracting a phenomenal 40 feet a day (the normal was 15) there was so little hope of anyone surviving that graves were dug so as to accommodate the bodies – that is if they were found and recovered.

Breaking into a gallery entrance, after *six-and-a-half days*, eleven corpses were found, victims of the aftermath of an explosion, their 'unharmed' condition indicating that each had succumbed to a slow suffocation in 'foul air'. The intrepid search team returned to the surface in order to allow the air, mixed with the smell of rotting flesh, to disperse. An unaccounted man was assumed lost, hidden by debris at the far end of the workings. On their return to the tunnel, to immense surprise, Sapper Bedson emerged, walking towards them – like an apparition – and uttered an extraordinary comment to the stunned searchers: 'For God's sake give me a drink! It's been a damned long shift!'. Bedson had lodged in a cavity within the face area, in total darkness. Years of previous mining experience and knowledge of how to survive if trapped after an accident kept him alive. Incredibly, two ration biscuits left in his pocket remained untouched and his bottle still contained drinkable water. To keep track of time he wound his pocket watch every 24 hours, having removed the glass so he could 'feel the time'. He also kept himself low, above the gas, sleeping when he could on a few sandbags.

Staffordshire-born, Bedson and his young family had settled in Conisbrough in about 1910, attracted by the prospect of work at Cadeby Colliery, one of two large and deep Doncaster pits owned by the Denaby and Cadeby Main Colliery Company. As an experienced hewer, William commanded a good wage, albeit in a far from safe pit. Early on the morning of 9 July 1912 an explosion ripped through the South District, killing thirty-five miners. The next day fifty-three rescue workers lost their lives in a second explosion. It is not known if Bedson was at work on the fateful night shift but probably had some role in the rescue operations, commanded by Hudspeth, a veteran of the Cadeby rescuers —referred to above. Bedson's experience in the Cadeby disaster helped him to survive one of the most extraordinary entrapments of the Great War.

It was at the relatively mature age of thirty-six that Bedson swapped the dark underground workings of Cadeby Main for the army. Along with a few of his workmates he joined one of the best-known local 'miners' regiments', the York and Lancasters, attesting on 13 November 1914. Posted to the Mediterranean eighth months later, he saw service in the Dardanelles – until 10 September 1915 – when he was wounded. In February 1916, fit again, he re-mustered into the Royal Engineers, initially as a tunneller's mate, but soon as a 'clay-kicking mole' – until his amazing rescue. On 15 July he was listed in the press as 'wounded', and then sent home for a deserved rest and recuperation. Sapper Bedson's army service continued, albeit at the Depot, until discharge on 13 December 1918. He got the Silver War Badge in respect of 'a gunshot wound to his left foot', but there was no award or mention of his outstanding courage and fortitude during the entombment of 7-10 June 1916.

After the war, Bedson recovered sufficiently so as to continue to work as a miner. Having to support his wife and four children and with only a small army pension he had little choice. In the 1939 Register, at the age of sixty, he is recorded as working as a colliery deputy, so must have done well in order to achieve this position. One wonders if he ever spoke to his family and mates about his entombment. Perhaps not, the knowledge passed down from other sources. William Henry Bedson, an 'everyday' miner but extraordinary veteran of the Great War passed away with little or no retirement, and devoid of commendations, in July 1945.

Wanted: veteran miners

In the few months leading up to Christmas 1915, through to January and even beyond the Military Service Act, the army's campaign to recruit experienced miners to the new tunnelling companies of the Royal Engineers continued unabated. It was the experienced 'front-line miners', particularly the hewers who got the coal using hand tools and/or basic machinery that were ideal for 'combing out' and 'fast-tracking' to the Western Front. Miners with twenty or more years of coal-face work, often over the age of forty, were welcomed into the ranks. Military training was of low priority for these veterans of the mines, their potential for 'sapping' and of course for their shaft and tunnelling abilities was far more important. As a further bonus they often had 'ambulance' and 'rescue' experience; and, always present, was their unselfish trait of coming to the help of a mate following a roof fall or explosion. An interesting case, probably repeated at recruiting offices and army depots elsewhere, occurred in the heart of the South Yorkshire coalfield in October 1915.

Samson Scargill's war

Six years married, with two small children, Samson Scargill left his small terrace house in Wombwell and walked a couple of miles to the recruiting office in the village of Darfield. It was a familiar route, not far from the big pit where he worked, Houghton Main. The sixth of October 1915 was a day he and his family would remember for the rest of their lives. A hewer, with well over twenty years of mining experience, he joined the Royal Engineers.

Samson's short service attestation form shows that he was forty-one years of age. His physical statistics: 5 feet 6 inches and just short of 11 stone may have been correct but his age was not as he was born in 1879. The discrepancy of five years was not accidental. Evidence of Samson's determination to join the engineers may help explain matters. Amongst his surviving military papers is a 'to whom it may concern' letter from his colliery manager, Mr S. Mellors, stating that he had left his employment 'on his own free will to join the force for Tunnelling, and would be a good man for the work named'. One can visualize Samson proudly showing the recruiting officer the testimonial, perfect for supporting his application. The 'age 41' may have been quite deliberately given by him or suggested by the official to demonstrate his maturity, and better secure acceptance.

Was joining the Royal Engineers a good move by Samson? Well it certainly benefitted the army. Within three weeks of 'signing his papers' he was on the Western Front, mustering as a 2s.2d-a-day tunneller's mate with the newly formed 252TC. The 252's commander Captain Rex Trower and his officers must have been impressed by Scargill, as he was appointed Lance Corporal on 23 May 1916 and promoted to Sergeant in April 1918.

In the spring of 1916, Samson was deployed in the Hebuterne-Beaumont-Hamel sector, digging tunnels, including the long gallery through the awkward chalk geology that reached under the German field fortification known as the Hawthorne Ridge Redoubt. During the operations the 252TC were supported by a large contingent (c.1,800) of infantry, such was the scale of the task. The huge mine, charged with 40,600lbs of ammonal, detonated under the Redoubt at 7.20 am on 1 July, completely destroying it.

Hawthorne Ridge mine explosion on 1 July 1916, filmed by Geoffrey Malins.

The spectacular event was soon to be seen by millions of cinema goers, in a film by the intrepid cinematographer Geoffrey Malins and his assistant John McDowell. It was to form the second part of the official war documentary *The Battle of the Somme*.[16] Malins, commissioned as a Lieutenant, described the scene that was witnessed in real time:

> The ground where I stood gave a mighty convulsion. It rocked and swayed. I gripped hold of my tripod to steady myself. Then, for all the world like a giant sponge, the earth rose in the air to the height of hundreds of feet. Higher and higher it rose, and with a horrible, grinding roar the earth fell back upon itself, leaving in its place a mountain of smoke.

Over four months later, on 13 November, at the start of the Battle of the Ancre, a second mine of similar magnitude was blown as part of the attack on Serre, lodged under the crater that had been created after the great blast of 1 July. It had been a long haul of somewhat frustrating sapping for Scargill and his unit, the 252TC having worked tirelessly for months creating thousands of feet of galleries, their spirit waning. But five days later the Somme Offensive was at an end. The experience gained from a coordinated use of infantry, artillery and tunnellers, provided the army with a much more effective force against the Germans when offensive mining was deployed. And the miner-tunnellers remained at the very heart of this achievement.

It was during sapping operations on the Somme front that Scargill was involved with acts of bravery so notable that he was awarded the Military Medal (MM), with an added clasp or Bar. Writing to his sister Ada in December, he confirmed his awards, though he had not yet been received them. He said that he had been actually recommended for honours by his superior officers on three previous occasions, winning the Bar for 'getting out some infantrymen who had been buried by the heavy shelling of the trenches' and that he 'got three out alive'. He explained, with typical modesty of so many similar awardees, that he had 'only done his duty' – 'the same as any other man would have done for me'. Samson also informed Ada that he had obtained a 'certificate of proficiency in the use of mine rescue apparatus', a further indication of his value.[17]

In November 1918 the Yorkshire Miners' Association (YMA) president presided over a ceremony in which Samson Scargill was presented with

Sampson Scargill. Brian Elliott

WOMBWELL MAN WINS MILITARY MEDAL.

Mrs. A. Dyson, of 136, Cuckstool terrace, Wombwell, has received the following letter from her brother, Lcp. Scargill (R.E.), who has been awarded the Military Medal. He writes: — "You ask me about the medal I won. I have not got the medal yet, but it is in the Army orders, and I am also awarded a bar as well. I was recommended three times

Sam Scargill's pit Houghton Main was a major single-colliery contributor of volunteers during the First World War. Brian Elliott

a gold watch on behalf of the owners and workmen of Houghton Main 'in recognition of his having been awarded the Military Medal for very gallant conduct in rescuing wounded soldiers under heavy fire, in France'.[18]

Samson Scargill returned to work at Houghton Main after his discharge, and was still 'going strong', working underground, aged sixty, as a 'packer' (stowing or packing loose stones and debris in waste areas, building walls at the end of old faces and supporting roadways), according to his entry in the 1939 Register. This remarkable war and pit veteran passed away at the age of eighty-three in 1963.

Sapper Hackett's ultimate sacrifice

Much has been written about William Hackett, the sole tunneller whose gallantry was recognised, sadly posthumously, with the award of a Victoria Cross.[19] William ('Bill') had difficulty in enlisting, turned down three times by his local infantry regiment, the York and Lancasters. An army-diagnosed heart condition was the major barrier to recruitment; and – at forty-one – he was regarded as too old for active service. However, later, as already noted above, the new tunneller companies not only allowed miners aged over forty to join their ranks, but actively encouraged their applications.

At last, on 'bonfire night', 1915, and now forty-two, Hackett enlisted. Within a couple of weeks he was on the Western Front, initially with the 172TC of the Royal Engineers, but soon attached to the new 254TC in the 'mining hot spot'

Sapper William Hackett.

of Givenchy les la Basse in northern France. Both companies had significant numbers of Welsh miners in their ranks, several of them featured in the VC action and rescue attempts outlined below.

In January 1916, William received the terrible news that his son, Arthur, aged fourteen, had had to have his right leg amputated following a haulage accident at Manvers Main. The boy had only been at the pit a month. Unable to read or write, Hackett was dependent on his tunneller friends to deal with correspondence, and with no leave granted it was a painful situation to endure.

Born in 1873, at Sneinton, Nottinghamshire, Bill Hackett found employment as a factory hand prior to taking the brave step of walking all the way into south Yorkshire and the distinctive pit community of Denaby Main, named after its local colliery. Still fresh in the minds of Denaby people were the cruel evictions of miners and their families from colliery company cottages during the so-called Bag Muck Strike of 1902-03,[20] but the prospect of 'good pay' was the magnet that pulled him and many others to Denaby.

Hackett worked at 'The Main' for twenty-three years, graduating from a coal-face 'filler' of 'tubs' (railed wagons) in 1901 to the elite position of hewer ten years later. Maybe homesick, in 1913 he vacated his tied 'pit house' and took his young family – wife Alice, and children Arthur (aged eleven) and Mary (nine) – back to Nottinghamshire. But another factor may have influenced his big decision. As mentioned above, Denaby's sister pit, Cadeby, had 'exploded' a year earlier, resulting in the deaths of more than eighty men and boys, including teams of rescue workers. However, within a few months, he was back in south Yorkshire, working underground as a 'dataller' ('day-wage repairer'), maintaining underground roadways, but on much lower pay. This was at the large Manvers complex, the Hackett family now residing in the small mining town of Mexborough. The disappointment of having to 'start again' as a lower-paid miner may have contributed to his abiding wish to leave his wife and young family to join the Colours, and 'escape from the dark'.

Hackett's unselfish action occurred on 23-24 June 1916, near Givenchy, at the 'Shaftsbury mine'. Accessed via a single shaft, a tunnel was being driven towards enemy lines when an explosion of a German mine collapsed about eight yards of the gallery, entombing Hackett and four other tunnellers. After two days of desperate digging, a rescue party managed to create a small hole which enabled William to assist three of his colleagues to safety, but he refused to abandon his seriously injured Swansea mate, 22-year-old Thomas Collins. Despite the further efforts of rescuers, including Welsh miner Albert Meadows, wearing Proto breathing apparatus, the gallery collapsed yet again, burying Hackett and Collins forever.

Sapper W.H. Vernon wrote a letter to Hackett's local alderman, who forwarded it to the *Rotherham Advertiser*.[21] This concluding excerpt concerns Vernon's eyewitness description of the attempted rescue operation and its aftermath:

> We kept them alive (Hackett and Collins) for three-and-a-half days, feeding them through pipes, but the Fates were again cruel. Another fall came, smashing the pipes. Having done our best, and knowing they had given their lives for their country, we retired to avenge their deaths. Our opportunity came with the morning. Fritz played one of his dirty tricks. Thinking to take us at a disadvantage, the Germans made a bombing raid, but they were surprised by our lads, and very few of them returned. We are up against it, but we go through it smiling.

A mix of official and spontaneous tributes to Hackett started to appear in the media not long after his death. His widow, Alice, was presented with his VC at Buckingham Palace in November 1916; and Hackett's colleagues from the Royal Engineers presented a gift of £67 to her in 1917, the cash used as an assistance for her children's schooling. A few months later, Alice was presented with an inscribed gold watch in a ceremony at Manvers Main.

Numerous monumental tributes relating to Hackett have appeared over the years, from the impressive Ploegsteert Memorial in Belgium to the new Tunneller's Memorial sited at Givenchy-les-la-Basse, France, unveiled in 2010. Locally, his family were especially proud of his inclusion on the Mexborough War Memorial, and a special Memorial Plaque was relocated to the site in 1997. Not far away, at the edge of Wath-upon-Dearne, Hackett's name is inscribed on the Manvers Main War Memorial, among more than 200 of his colliery workmates; and 'at home', in the grounds of Nottingham Castle, the Nottingham and Nottinghamshire Victoria Cross Memorial was unveiled a few weeks prior to the opening and dedication ceremony of the Tunnellers' Memorial in Givenchy, referred to above.

It's hard to find suitable words to describe the action of William Hackett, who preferred to help an injured mate rather than escape with his own life. For him there was no choice in his decision, the welfare of a fellow 'miner-tunneller' was *the* priority. Though during mining accidents and disasters this kind of action were not uncommon, his Victoria Cross must rank as one of the

Ploegsteert Memorial to the Missing (Berks Cemetery Extension), Belgium. The impressive memorial is located 12km from Ypres but accommodates names of missing soldiers in various actions south of the Ypres Salient. Hackett's name is inscribed under a 'Sapper' heading on the Ploegsteert Memorial. Brian Elliott

Local military historian Andrew Featherstone points to Hackett's name on the Manvers Main Colliery Memorial, near Wath-upon-Dearne, Rotherham. Brian Elliott

most worthy of all awards given in the Great War. By all accounts a quiet and unassuming character, Hackett really was one of the bravest of the brave.

High Wood: 'the rottenest place on the Western Front'

William Sneddon Wilson, aged twenty-three, a former Scottish miner serving as a Lance-Corporal in A Company of the 10/11 Highland Light Infantry (HLI) lost his life at the 'Switch Line', on elevated land close to the northern apex of High Wood, during the Somme Offensive, on 17 August 1916.[22] The Welsh miner-soldier and author Frank Richards provides a glimpse of the horrible conditions in and around these trenches about the time of Wilson's death:

> Some parts of the parapet had been built up with dead men, and here and there arms and legs were protruding. In one bay only the arms and heads of two men; their teeth ... grinning horribly down on us ... Great trees were uprooted and split like matchwood, some falling over the trench. We were throwing our dead on the back of the parapet, from where in some cases they were blown up again and thrown further afield.[23]

Much of the 75-acre wood had transformed into a ghoulish and hideous spectre of tree stumps and bomb-blasted ground. Wilson's body, like many thousands of other soldiers killed or abandoned here was never found, lost amid an anonymous mass grave.

A month before Wilson's death, Private Thomas Lyon of the 1/9 HLI described the 'greeting' that his pals got during one of numerous thwarted assaults on High Wood, when the air was 'full of flying, hissing bullets': '... a withering blast of machine-gun fire met them at once, and many stumbled and fell ere they had gone more than a few yards.'[24]

When the battalion mustered on the morning of 16 July only 200 'answered

Portrait newspaper image of Corporal William S. Wilson alongside his 1914-15 Star medal. Brian Elliott

Detail of the reverse of Wilson's 1914-15 Star medal.

1916: Unspeakable Horror

the first roll call', out of about 800 who had 'gone forward'.[25] The wood was eventually overcome on 15 September 1916, a month after Wilson's death. Over sixty-four advance and retreat days, British battalions had used a variety weapons and strategies, from machine guns and giant flamethrowers, to cavalry charges and offensive tunnelling. A few tanks were also deployed.

After battalion training, Wilson had embarked for France, landing with the 10th HLI at Boulogne on 12 May 1915, but was sent home for a short period of recuperation just four months later, suffering from a gas attack, at Loos.

William's death must have been a terrible blow for his parents who may have felt a good deal of relief when the young miner left Newbattle collieries for service with the HLI a few days after the start of the war. His employers were the Lothian Coal Company whose showpiece Lady Victoria Colliery helped change Newtongrange into a large colliery settlement. Residing in a company house, with union involvement outlawed, the Wilsons were at the mercy of owners and managers regarded as some of the most strict in the Scottish coalfields.[26] Initially classed as missing, his death took almost a year to become officially recognised. Like many of his pit mates, he had escaped from an authoritarian regime only to lose his life at the 'rottenest place on the Western Front'.[27] William S. Wilson is commemorated on the Thiepval Memorial to the Missing of the Somme (Pier & Face15C), and locally on memorials at his home church and village, in Newtongrange, Midlothian.

Earlier, on 4 August 1916, close to High Wood, 28-year-old machine-gunner John Thomas Thornley, a miner from Greenside Colliery, Ryton, near Newcastle upon Tyne lost his life in the most unfortunate of circumstances. The (101st) Company War Diary includes reference to the situation that took place at Bazentin-le Petite: 'Nos.10 and 11 Guns did more good work on enemy, trying to reinforce over the open to the left and in their sap out from High Wood. These teams suffered 3 casualties (1 killed and 2 wounded) from our own artillery whose shells were falling short.' Losing his life in friendly fire must have been a massive blow to endure by his widow Isabella who was left to care for three small children, and her husband's body never recovered. For the wider Thornley family it was a double tragedy, John's younger brother James, aged twenty-one, serving in the 14/Durham Light Infantry, losing his life a few months earlier, on 19 April 1916.

Senghenydd to Somme: 'Reg' Lambert's epic last journey

The Lambert family, comprising Edwin, his wife Alice and their six children, had settled in Senghenydd in south Wales by about 1910, moving there from the Forest of Dean, where mining was an ancient occupation and way of life. Edwin Reginald Lambert, aged fourteen and William Lambert, fifteen, are recorded as 'colliers' in the 1911 census, probably working as pony drivers. Their father Edwin had obtained employment as an 'ostler', in charge of the horses in part the big pit at Senghenydd, the Universal Colliery. The move to the Aber Valley was perhaps an understandable outcome for the family as Edwin's wife Alice was Monmouth-born. But it was better pay that was the major motivational factor for their migration. Within twenty years Senghennydd had expanded from a population of a few hundred to over 6,000, mostly incomers attracted to the Universal. But it was a far from safe place to be employed. Eighty-one men and boys lost their lives there in a great explosion only ten years earlier. The Lamberts, Edwin and sons Reginald and William, had a fortuitous escape from death themselves, when the pit 'fired' again, with a catastrophic outcome, on 14 October 1913, 439 men, boys and a rescuer losing their lives, in the worst mine disaster in British coalmining history.

'Reg' Lambert, detail from a studio photograph probably taken just prior to enlistment.
Laurin Espie/Arnold Kingston: findagrave.com/memorial/12458907/edwin-reginald-lambert; everyoneremembered.org (Royal British Legion)

Senghenydd was absolutely devastated for many months, so on 3 September 1914 Reginald, along with his brother William, walked to the recruiting office at nearby Abertridwr. Declared as nineteen years old (he was a little younger), Reg was 'graded' as 5 feet 6 inches in height and weighed 129 pounds, but was welcomed into the 'Hussars of the Line'. His equine experience may well have been an advantage. But he was soon transferred to the 2nd York and Lancaster infantry regiment, well known for its recruitment of miners. The shafts at the Universal were even named 'York' and 'Lancaster'.

On the Western Front in the Ypres Salient from July 1915, by early 1916 Reg had contacted the contagious skin disease impetigo. He recovered quickly and must have impressed his superiors since he got promoted to Lance Corporal, albeit on an unpaid basis, on 26 May. Just over a year later, on 25 September, Reg was active on the Somme front, as part of the Fourth Army's advance against German trenches and positions, on an eventful first day and night of what became the four-day Battle of Morval. The 2nd York and Lancasters were relieved the next day but Reg Lambert was already reported as 'wounded in action' and then – worryingly – was declared 'missing'. With no news forthcoming, Alice was desperate for news of her son, writing to the regiment on 18 October 1916. His remains were lost in the fields of Morval. Lance Corporal Edwin Reginald Lambert is commemorated on the Thiepval Memorial to the Missing (Pier and Face 14A and 14B) and is one of sixty-three Great War names honoured on the Senghenydd village war memorial.[28] Like numerous other young Senghenydd military recruits, Reg Lambert had escaped death in a terrible pit disaster only to perish on the bloody landscape of the Somme.

Under the Somme at Bouzincourt

In 2016, as part of a 'tunnellers'' battlefield tour, I was able to descend – via a series of spiral steps – 40 feet or so (c.4 metres) below the church of St Honore, in the small village of Bouzincourt, northern France.[29] Here, from 1916, sixteenth-century caves were enlarged, refashioned and extended by British tunnellers into safe and secure resting places, creating a candle-lit refuge for weary Somme soldiers. Located just behind the British lines for most of the war, Bouzincourt lies about four miles north-west of Albert.

The geology is chalk, perfect for marks by pencil or a small blade. Torch light reveals an astonishing range of graffiti inscribed with thought and care by British, Canadian and Australian soldiers. From names and regiments to witty messages, poems and drawings, the images certainly connected with me; and perhaps more so for the increasing number of descendants, now able to view them either directly or by remote access to a database compiled by the village's historical society.[30] About 2,100 examples have now been recorded for the Bouzincourt site. Also on view were a variety of artefacts: abandoned boots, helmets, wheelbarrows and excavation tools, even bullets and rifles.

Exploring the tunnels of Bouzincourt was one of the most amazing experiences imaginable, made even more emotive when, later, I stopped to think about the rationale of the soldiers' intentions. The markings were

Graffiti on the excavated cave walls. Brian Elliott

far more than random doodles, but served as potential last messages to loved ones; and maybe a record to be discovered generations later. To many, the inscriptions provide a direct connection to the ordinary soldier in the First World War.

The Bouzincourt 'muches' remain the most compelling of many subterranean First World War sites where soldiers made inscriptions, at least 400 identified by battlefield archaeologists and researchers in the Picardy and Nord-Pas-Calais regions alone. The 'Underground City' museum at Nours is where the public can view several thousand graffiti items, many of them inscribed by Australian soldiers.[31]

The old church of St Honore in Bouzincourt was but a ruin by the end of the war, but rebuilt above its secret labyrinth of caves and tunnels in 1920. Its architect and builders created a landmark tower in the form of an artillery shell, fronting the village's war memorial.

Bouzincourt church and village war memorial. Brian Elliott

More 'miner-VCs'

Four former miners (and the son of a colliery manager) were recipients of the Victoria Cross during 1916, a relatively low number in view of the volume and extent of their service at this time. But the secrecy of the subterranean war on the Western Front may have limited award recommendations. The great exception was the posthumous VC awarded to **Sapper Hackett**, already referred to on pp.127–130. Two Welsh colliers and a Yorkshire colliery office worker complete the tally, getting their award in the spring and summer months, from April to July.

It would be remiss, however, not to mention **Lieutenant Colonel Roland Boys Bradford**, commander of 1st Battalion of the Durham Light Infantry, who was awarded a VC for his actions and inspired leadership – despite being wounded – at Eaucourt L'Abbaye, France on 1 October. Sadly, Bradford was to lose his life on 30 November 1917 at Bourlon Wood. Remarkably, his older brother, **George Nicholson Bradford**, a Lieutenant Commander in the Royal Navy, was also awarded a posthumous VC for an attack on Zeebrugge, Belgium on St George's Day, 23 April 1918. They were the only brothers to get VCs in the

1916: Unspeakable Horror • **135**

Great War. The mining connection was that their father George Bradford was a well-known mining engineer and colliery manager from Bishop Auckland, County Durham, who went on to take charge of a mining complex in south Wales. It's known that Roland Bradford in particular commanded the respect of his men, and was always aware of their welfare, emulating his father's style of management.[32]

Private James ('Jim') Henry Finn (aka Fynn)[33] was of Cornish heritage, born in the workhouse at Truro, from a large family residing in a Bodmin lodging house. He had no formal education, but his family's military background was impressive. Jim's father, Frederick, was a Boer War veteran, serving in the Duke of Cornwall's Light Infantry (DCLI); three brothers were in the Royal Navy and another also in the DCLI. Two of his five sisters were military nurses. Finn enlisted in the DCLI (5th Battalion Territorials) but 'evangelical religion', in particular the Salvation Army, became a prominent part of his life. After a short period working in London, he moved to Monmouthshire, obtaining a job at Cwmtillery Colliery[34] and lodged with a pit-pal's family, at Abertillery. In south Wales he continued his Salvation Army interests, as a patrol leader and, via the Scouts, was able to learn and practise first aid and lifesaving, his troop winning a prestigious national competition held at the Crystal Palace. Such skills made him a useful 'military medic' in trenches and on battlefields.

Jim Finn volunteered for military service shortly after the outbreak of the War, preferring to join the local infantry regiment, the South Wales Borderers, rather than more generic RAMC, which may have suited his Salvation and scouting background far more. In France with the 4th Borderers from the spring of 1915, Finn combined bravery with his sense of fellowship when, under heavy fire, he dashed to rescue his commanding officer, who, sadly, was either already dead or died shortly afterwards, according to a local newspaper report.[35] Not long afterwards, he was badly injured (bullet wound to the knee and a shrapnel wound in the chest) when the trench he was occupying with four mates received a direct hit, burying them under debris, with Jim the sole survivor. Nevertheless, he was still able to rejoin the 4th Borderers, at Gallipoli, for their Mesopotamian campaign.

The extreme act of bravery that resulted in Finn being recommended for a VC took place at Sanna-I-Yat (during the second battle there), Mesopotamia (Iraq) on 9 April 1916. In a lone forward position, dug-in about 300 yards from an enemy trench, he ignored continuous heavy fire in order to make several precarious journeys to bandage several injured soldiers. Unable to get a stretcher, he carried one of the injured on his back to safety and returned to get another in similar circumstances. Jim Finn remained at the Mesopotamian front but on 29 March 1917, ten months after his VC action, at Marl Plain, he sustained a leg wound and whilst being stretchered to the field ambulance was shot yet again, dying the next day. He was twenty-three years old. Finn's VC was presented to his father by George V, in Hyde Park, London on 2 June 1917.

The former Cwmtillery miner, a dedicated Christian from a modest background, was totally unselfish in his actions, ultimately sacrificing his own life. Although aware of his supreme accolade, it was tragic that he was unable to attend and receive his Victoria Cross investiture.

Joseph John Davies, a corporal (later sergeant) in the 10th (Service) Battalion of The Royal Welch Fusiliers, son of a Royal Fusilier veteran and Staffordshire coalminer, got his VC for his actions prior to an attack on Delville Wood. On 20 July 1916, when he was with a section of eight men from D company, the group were surrounded in a second counter-attack by the Germans. Davies led the men into a shell hole and through bomb throwing and rapid fire not only managed to repel the enemy but chased and bayoneting several during their retreat. On the same day, but in a separate action, a fellow Welsh Fusilier, Private Albert Hill, also got a VC, a remarkable outcome for the 10th Battalion. Prior to enlistment, Hill worked as an apprentice hat-maker, but his late father was a Lancashire miner, at Ashton Moss Colliery.[36]

Staffordshire-born, from Tipton, Joseph Davies worked as a teenage miner prior to enlisting in the Welsh Regiment in 1909, so was already a seasoned infantryman before the start of the Great War, serving in Egypt and India. Davies had experienced a number of wounds in 1915, one at the Second Battle of Ypres and another during a German trench raid, the injuries entailing an 'unfit for service' assessment and transfer to Blighty. It was after medical reclassification that he returned to overseas service, initially in Gibraltar, in May 1916. He was classed as unfit for service yet again, this time permanently, probably after sustaining a serious shoulder wound. Pictures of him in October 1916 at his VC investiture show him with his left arm in a sling. His injuries certainly precluded a return to his pre-service employment of mining. Davies spent his later life in Dorset, passing away in 1976.

Private G.W. Chafer VC. Public domain

Less than a month before the start of the Somme Offensive, **Private George William ('Willie') Chafer**, serving in the 1st East Yorkshires, was awarded the VC for his 'conspicuous bravery' when delivering a message carried by a 'collapsed' colleague, near Meaulte, on 3-4 June 1916. His action took place even though he was already wounded in three places, his left leg having to be amputated in hospital later. Bradford-born but brought up at Epworth, Lincolnshire, 'Wee Willie' previously worked as a weigh clerk at Silverwood Colliery,

Rotherham (see pp.82–105), and was only accepted by his chosen regiment on his second attempt to enlist, apparently due to his small stature and 'frailty'. Chafer's Victoria Cross presentation – along with his background and his action – was celebrated in considerable detail by his local newspaper, the *Rotherham Advertiser*.[37]

***

It's hardly surprising that for many of the former miners who served in the armed forces in the Great War, the year 1916 was one to try and forget rather than remember. So many were lost on the Somme and countless others had to endure physical and mental wounds for the rest of their lives. No wonder many miner war veterans refrained or refused to speak about the terrible conditions they endured and the horrors witnessed. Thankfully, some military historians have had the foresight to interview many First World War veterans, recording testimonies from men close to the end of their lives, many speaking 'in the public domain' for the first time.[38] We owe great thanks too, to family and local historians who have recorded and recounted individual stories and shared these for others to see on online platforms such as the Imperial War Museum's Lives of the First World War.[39]

Personally, I will never forget walking through the terraced streets of the old pit village of Hoyland Common, near Barnsley, in the company of an old miner and local historian in his late eighties.[40] Every few paces he stopped to point at front doors, where families had lost one or more of their household during the Great War. He could name them all. A single terraced street, Wood View, in the adjoining rural community of Tankersley, lost one (or more) inhabitants from every house.[41] Not far away in the pit village of Darfield, 82% (111 men) of known servicemen killed in action were miners, fifty-one of them (46%) from one colliery – Houghton Main – prior to enlistment. Thirty-two of the men killed in action formed a compact group living in the heart of the old village, a dozen of them casualties in 1916.[42] It was akin to the prolonged aftermath of a pit disaster, unimaginable grief strung out over several months, some bodies identified but others missing, left in the ground.

Similar patterns of loss and bereavement must have been repeated in other close-knit mining communities of Britain. It was mainly the younger miners who had volunteered with their mates in 1914-15, usually from the same pit, who found themselves on the Western Front from late 1915 or early 1916. The average age at death of the Darfield cohort referred to above was 26.9 years. A few of the more mature volunteers, however, had inflated an otherwise very young group, some 'hiding' their true age so as to ensure enlistment. Private Thomas Conway, of an Irish heritage, who had enlisted as a 'Barnsley Pal' (13th Battalion York and Lancasters), on 1 June 1915, gave his age as thirty-six years and five months but was at least ten years older when he lost his life on the first day of the Somme battle, his remains never recovered. As we have seen, many mature hewers were 'combed out' of their pits in 1916, and rushed to the front to work as tunnellers' mates and tunnellers.

Tribunals and Schemes

From the spring of 1916, despite the Military Service Act, many 'miners' (pit-top workers and conscientious objectors ['COs'] especially) continued to face a variety of tribunals and government schemes regarding their eligibility for service in the military. For many applicants, a bewildering shebang of bureaucracy in the new courts and tribunals was hard to comprehend and inconsistencies between them did not help matters. The main ones are summarised below:

Colliery Recruitment Courts (CRCs)

The CRCs appraised applications for exemption from miners 'barred' or 'not barred' by the Military Service Act. 'Pit-top' and colliery office workers, for example might apply for exemption, though were unlikely to succeed. These courts also heard evidence from men who had earlier attested under the Derby Scheme.

Twenty-three CRCs served the coalfields in 1916, presided over by a Divisional Inspector of Mines, alongside representatives from the military, coal owners and miners. They were submerged with work due to a great rush of applicants. In 1916, 85,000 'certificates of exemption' were issued by the Northern Division alone, according to mines inspector J.R.R. Wilson in his annual report. By the end of 1917, J. Dyer Lewis, the Welsh Divisional Inspector of Mines, who chaired courts in Cardiff, Swansea and Newport, referred to

The Military Service Act (1916-18)

All single men (and widowers with no dependant children) aged eighteen to forty-one and residing in Great Britain on 15 August, were compelled to some form of military service, operational on 2 March 1916. Application for exception including 'conscientious' claims were decided via tribunals. The Act was extended to married men on 25 May 1916 and the upper age was increased to fifty-one in 1918. Most miners remained exempt from conscription because of the importance of the coal industry to the war. However, under the new Act exempted or 'starred' miners only referred to those men who worked underground or in certain pit-top occupations such as winding engine man, pump man, weighman, electrician, fitter or mechanic. Conscription ceased on 11 November 1918 (although the Act remained on the statute books until 1927), and any former miners still in service were discharged on 31 March 1920.

his and his colleagues' workload as 'overwhelming'. Not surprisingly, errors of judgement were not uncommon.

The overseeing body of the CRCs, the **Central Colliery Recruitment Court**, at its first meeting in January 1916 decided that 'suitable miners' could be spared for the new tunnelling companies, and agreed that 10,000 'skilled miners' should accordingly be enlisted each year.

Military Service Tribunals (MSTs)
Individuals and employers who wished to appeal against compulsory military service had to do so by attending a Military Service Tribunal (MST), unless a certificate of exemption was issued earlier (also see Colliery Recruitment Courts above). More than 2,000 local MSTs were established – staffed mostly by local councillors – to hear objections to conscription, with the possibility of further challenges via Regional Appeal Tribunals and a Central Appeal Tribunal. The categories to be evaluated for exempted status were limited to 'work of national interest', ill-health or infirmity, extreme hardship and (the hardest to interpret) conscientious objection (see Chapter 6). Service involving exclusion from combat was a more likely outcome for COs' appeals, as was 'conditional exemption', which usually meant classification within the 'Work of National Importance' category (see 'Brace Scheme' below). As you will see it was a complex situation.

The Pelham Committee ('Work of National Importance' Scheme)
Appointed by the Board of Trade on 28 March 1916, under the chairmanship of T.H.W. Pelham, the committee advised the MSTs of the type of employment that was acceptable *as an alternative to military service* for conscientious objectors. Its preliminary list of appropriate occupations included work in mines. By 10 May, 265 cases had been referred to Pelham for interpretation and by end of hostilities total referrals almost reached four thousand (according to Walter Runciman, President Board of Trade, questioned in the House of Commons). Christadelphians formed the largest group, at around 43%. Only one in three of all referrals were actually exempted.

Committee on the Employment of Conscientious Objectors (aka 'Brace Scheme')
This Government scheme commenced in June 1916, under the leadership of the Under-secretary of the Home Office, the South Glamorgan MP William Brace. It was started so as to allow COs to undertake 'work of national importance' rather than remain in prison. Brace was chosen because of his previous experience as president of the South Wales Miners Federation. Hubs known as Work Centres appeared and dreaded 'tented camps' were established in harsh, out-of-the way locations. Some, like Princetown, Dartmoor, became notorious for the brutal and soul-destroying tasks allocated to internees.

Non-Combatant Corps (NCC)
The NCC was created in March 1916 as an army unit for COs, its 'volunteers' not needed to fight or take part in military combat. The CO members, all privates, had to wear military uniform and confirm to military discipline. Receiving less pay than other soldiers, they were used for mundane labouring tasks and referred, disparagingly, as the 'No Courage Corps'. Referrals who refused to join were court-martialled. Both the No Conscription Fellowship and the Society of Friends (Quakers) saw the NCC as part of the state's 'military machine'.

Aftermath

The fallout from the Somme Offensive reverberated through the coalfield communities of Britain like the aftershocks of a great earthquake, when thousands of discharged 'miner-soldiers' returned home suffering from a variety of physical and mental conditions. Despite the demand for coal, most were unfit to resume work at their old jobs, especially if they were coalface men, former hewers whose skills and physical attributes were so vital for getting the coal. Yet some, even a few amputees, did return, working in or above their mines for many years afterwards. Thankfully, some examples of these otherwise faded and forgotten heroes of the war have been researched and recorded by family historians, recognizing the huge contributions that they made. The courage of the survivors in returning to the mines – despite their condition – was extraordinary. The following example, taken from my home area of Barnsley, was by no means unique. They just got on with what pit job they could get. There was little choice.

Twenty-year-old **John William Exley**,[43] a former pony driver who probably worked underground alongside his father Charles, a hewer, at Monk Bretton Colliery, enlisted with the 5th York and Lancasters during the first great wave of miner-volunteering. In France from 12 April 1915, Exley's overseas service only lasted until 13 July 1915, ninety-two days. This was due to a serious wound. Dispatched home for urgent medical attention, his left arm was amputated at 'Epsom hospital', according to family knowledge. When on active service with the 5th, he was involved in trench warfare within the Ypres Salient, and may have participated in the Battle of Aubers Ridge on 9 May. Exley was one of the many casualties that his battalion suffered in the spring and early summer of 1915, a situation much accelerated at the start of the Somme Offensive.

Only a few weeks after his hospitalisation, towards the end of 1915, in St John the Baptist's church, Barnsley, Exley married Harriet Lowe whose father Noah also served in the York and Lancasters (15th Reserve Battalion). He remained on home service until his army discharge (he was classed as 'physically unfit') on 5 May 1916. In a remarkable turnaround, John returned to work as a miner, working underground 'on the haulage' – as 'rope runner' – normally an arduous, two-handed job. Furthermore, he played football to a high standard for a local team, winning a gold winner's medal to go alongside his three Great War medals (Victory, British and 15 Star). Twenty years later, according to the 1939 Register, he was still at work albeit on lighter duties as a night watchman. His son, Charles, carried on the family mining tradition, as a colliery haulage hand. John William Exley, Great War veteran and one-armed miner and footballer, died in Barnsley in 1961, aged sixty-seven, after an eventful life.

Another forgotten, overlooked cohort were numerous veteran miners – rushed to the front in early 1916 – who suffered so badly whilst engaged in extreme tunnelling operations that they had to be sent home, with limited prospects, after only a few months of active service. Here are a couple of examples, the more startling in view of the fact they were men well used to mining in low, fiery and uncomfortable seams in Staffordshire.

Richard Howe, a veteran hewer at Fenton Glebe Colliery (born at Stoke-upon-Trent in 1868), enlisted in the Royal Engineers at the age of forty-seven, on 3 September 1915. Within ten days he had embarked for France. By early March 1916, after six gruelling months with the 175 tunnellers Richard was discharged from armed service under the 'Para 392' rules, as 'not being likely to become an efficient soldier'. Myalgia was his principal ailment, worn-out muscles no doubt due to repetitive use of shovel and picks whilst acting as a human mole in the most horrible conditions imaginable. Sent home and placed in a Provisional Corps (holding unit) of the Royal Engineers, he was allocated a Silver War Badge, alongside with a tranche of other former tunneller-miners, in May 1917. Linda, his wife, was very happy and relieved at his homecoming but not at the reality, thinking if her man would find sustainable employment ever again.

William Smallman enlisted at the age of forty-five according to his attestation form but was actually fifty-three years old (born at Dudley in 1862). The head of a large family, this Cannock collier had a lot to offer when he joined the Royal Engineers, in particular the 178 Tunnelling Company, as a '2s.2d a day tunneller's mate', on 10 August 1915. Like Howe, he was in France within days, his background sufficient for a re-mustering as a full-blown tunneller by October; but the Sapper's service barely extended into 1916, as he was discharged, also

John William Exley, as one-armed footballer.
Exley family history

Sapper Richard Howe's Victory medal.
Brian Elliott

under 'para 392', on 21 January. Hospitalised 'in the field' and then 'at home', the chances of mine work for the now fifty-five year old must have been remote.

Howe appears to have brushed off his disabilities, or at least put up with them, finding a job at nearby Mossfield Colliery, Longton by 1920. Almost ten years later, residing at Fenton, he was still working – albeit as a general labourer – though the 1939 Register does not state that this was at a colliery, passing away at the age of seventy-eight in 1946. The future of Smallman is not known.

Light Lines

In 2016, an acclaimed artwork – Light Lines – was displayed along the front of Barnsley's town hall to commemorate the c.300 local men who were killed on the Battle of the Somme's most fateful of days: 1 July 1916. It consisted of a series of thirty acrylic panels, laser etched with soldiers' faces, some identified, others shown in outline. Designed and created by Neil Musson and Jono Retallick (musson+retallick), the installation, which includes plaster medals made by local schoolchildren, changes in appearance according to the light and at nightfall is even more dramatic due to the placement of LED lights at the base of each panel. The impact is both stunning and compelling. The whole project was impossible to complete but for the research of local people interested in their own and their friends' family histories. The installation was praised for the way in which the artists and commissioner (BarnsleyMBC) had specified and harnessed public participation – via schools and communities – in its making and display. A Somme exhibition in the Town Hall's Experience Barnsley museum was also a most meaningful tribute, attracting many visitors.[44]

The 'Somme Exhibition' in Experience Barnsley Museum (24 August-20 November 2016) was of course one of many similar ones held in former coalmining towns and villages of Britain. Many illustrative display boards and personal items of memorabilia caught my eye when viewing in Barnsley but one example really did stand out: Fred, Charles and Ernest Walker, brothers-in-arms with the 1st Barnsley Pals (13th York & Lancasters) were killed in action on 1 July. Goodness knows how many other brothers had the same fate, collier lads from the same pit who volunteered amid the rush to enlist after the start of the war in 1914.

The acclaimed Light Lines art installation in Barnsley, 2016. It was re-installed in later years.
Brian Elliott

Chapter Six

'CONCHIES':
THE MEN WHO SAID NO

Although relatively few of the total number of miners in employment appear to have objected to military service on grounds of 'conscience', they formed a small but significant proportion of conscientious objectors (COs) in many pit villages, especially in areas with radical traditions. Philip Adams has shown that Port Talbot had over 70 COs and at least one in five of these were miners, a situation that was probably typical of the South Wales Coalfield as a whole.[1] Adams also points out that there were many miners who were not COs but offered great resistance to military service due to their political and pacifist views. Among these were Henry Davies of Cwmafan, where he was a leading ILP figure. Born

Conscientious objectors at Princetown, Dartmoor in c.1917. Brian Elliott

in 1870, so too old for conscription, this charismatic former miner worked tirelessly to represent the COs, armed with nothing more offensive than a detailed knowledge of the Military Service Act.[2] A close ally and of equal impact was Davies's friend Tal Mainwaring, a 'badged' miner, ILP and NCF member, who in June 1916 was imprisoned under the Defence of the Realm Act (DoRA) for anti-conscription activities.[3]

There is evidence to suggest that the COs and the pacifists were strongly opposed and disliked by some of their fellow miners. In County Durham, Thomas Cann, the general secretary and leader of the DMA openly condemned 'their stand', threatening that 'if they did not toe the line ... more strenuous steps should be taken'.[4] This antagonism was certainly apparent in parts of south Yorkshire, union members at Manvers and Silverwood collieries reported as 'strongly expressing their resentment' of 'pacifist elements in their delegates'; and YMA leader James Walton (elected MP for Don Valley at the 1918 general election) told the Manvers miners that as many as three-quarters of the delegates to the Association were 'pacifists and Bolsheviks'.[5] The Manvers men had refused to work with a union official 'sheltered from military service' unless he joined the army.

At a local union meeting on 18 March 1917, the miners at Hebburn Colliery, near Jarrow, took the extraordinary decision not only to exclude two of their members but to refuse to work with them. The 'barred' men, Richard Main, aged thirty-nine, a local councillor, and 25-year-old John Richard Cable, described as 'Quakers and Socialists' and 'conscientious objectors', had turned down an opportunity to transfer to a Work Centre under the Brace scheme but were still prepared to work as miners. The decision of the Hebburn Lodge was backed by the pit's manager, who offered the two 'offenders' a fortnight's wages in lieu of the termination of their employment. Main and Cable appealed to the DMA against their sacking. The Executive not only ruled in their favour, but condemned their treatment by the Lodge, withdrawing a resolution of a county-wide ballot as to whether COs should be union members. A further appeal to the DMA by the Lodge on 30 June was again turned down by the executive committee. Despite this, the miners at Hebburn continued to refuse to work alongside Main and Cable, who were arrested on 18 July. Their subsequent appearance at the magistrates' court (where they were fined 40 shillings and 'handed over to the military') and their court martials were widely reported in local and regional newspapers. Both men were condemned to prison with hard labour, initially in Wormwood Scrubs (and then Armley, Leeds, and in the case of Main, Liverpool prison too); and were not discharged until March (Main) and April (Cable) 1919, after serving eighteen months 'behind bars'.[6]

Those CO miners who had the courage to return to their pits after the war were often 'cold-shouldered' by some of their workmates. An even more extreme attitude took place at Whitburn Colliery, near Sunderland. Here the pit was described as 'idle' on 6 May 1919 as the men had 'downed tools' and refused to work with two conscientious objectors. The pair of 'offenders' had in fact been accepted back to work by local union officials, but the latter were regarded as 'known sympathisers of the conscientious objectors' movement' by the men.[7] About the same time, at nearby Silksworth Colliery, the miners refused to work with a returning CO, George Wright[8].

A very sad case occurred in the County Durham pit village of Quebec concerning a young miner, John George Winter, a Methodist and Socialist who refused to attest or answer any tribunal questions. Court-martialled at Newcastle on 5 July 1917, the 25-year-old was to spend only a few days of his sentence in Wormwood Scrubs, succumbing to double pneumonia on 13 July 1918. An unconfirmed though convincing cell-mate account of his demise referred to a beating that Winter had had from the warders as the real cause of his death. At the funeral, hatred from some spectators ran so high that stones were thrown at his coffin; and at the graveside 'boos and the singing of the *Red Flag*' ruined what should have been solemn and peaceful ceremony for the Winter family.[9]

Cable, Main and Winter are just three of at least 165 miners recorded in the Pearce Register of British Conscientious Objectors (PRBCO), by far the most comprehensive source for discovering the names, backgrounds, motivations and treatment of First World War COs. Although this figure represents only about 1% of the 17,426 names listed in 2017, and is inclusive of a small number of miners' sons, former miners and trade union officers, many of the individual entries are sufficiently detailed as to enable analysis of the whole data; and also provide an insight into the lives of the miners who tried to resist conscription through religious, pacifist and political reasons, especially during and after 1916.[10]

In occupational terms, most (82%) of the 'miners' identified in the PRBCO database are referred to generically, meaning that some were surface workers, therefore excluded from 'reserved' status under the Military Service Act. However, some 'exempted' men – underground hewers, haulage workers and craftsmen – are listed, so were certainly entitled to continue working.

Some COs were so determined to air their beliefs in public that they applied to tribunals for exemption – even though they were of 'starred' status. The *Colliery Guardian* reported that miner COs in Northumberland and County Durham were insisting on appealing against military service so that they could air their objections to serving in public – despite the fact they were exempted.[11] Three months later, the *Durham Chronicle* reported that an (unnamed) Auton Stile miner applied for exemption 'on the grounds of religious objection' even though he worked as a hewer. The chairman of the rural tribunal dismissed the man's motives as 'balderdash'. Even a supportive letter from the miner's colliery manager, saying that the applicant would be 'missed very much', had no influence. Arrangements were made for him to 'join the army for non-combatant service', via the RAMC.[12]

At times the Tribunals and Tribunal Appeal Courts were extremely suspicious of the 'high and mighty motives' of any miner who appeared before them. Scottish miners in particular appear have been treated more harshly than others. In January 1917, an unnamed Kirkintilloch man (formerly a grocer's shop manager) appeared before the Dumbartonshire Appeal Tribunal, contesting his 'call to the Colours' as he was a conscientious objector who had now found 'work of national importance' (via the Pelham Scheme) in a coalmine. The Tribunal dismissed his appeal on the grounds that his mining occupation was a very recent one, but little or no weight was given to his beliefs.[13] John Lister, a 22-year-old miner from Kelty (Cupar, Fife) gave a 'long tirade' of a speech against the war and against capitalism but his appeal was summarily dismissed due to his 'radical' stance.[14] Four young miners from the Leven area: John Paton (aged nineteen), William Dick (twenty-four), John Hastie (twenty-eight) and William Carrie (twenty-eight), had their appeals not to serve on CO grounds summarily dismissed at the Leven and Brockhouses Tribunal.[15]

In Lancashire, a 39-year-old coalminer appealed for exemption at the Burnley Tribunal on conscientious grounds even before he had been 'combed-out' (by ballot). After confirming his 'Christian religious beliefs' the chairman asked him if he was 'the only Christian in Burnley?', to the great 'laughter' of the assembly, a kind of humiliation that was not uncommon. His case was adjourned 'to see if he was to be combed-out'.[16]

Consistency in the interpretation of the regulations concerning miners and 'recruitment' to the military varied from one tribunal to another and certainly did across the coalfield regions. The situation was, however, quite complex after amendments and updates were introduced. In households where two-thirds of the 'family' were of 'military age' – and already recruited/combed-out – any remaining working miner should be exempted. Some Tribunals stretched the criteria to 50%. Discretion and common sense, however, did prevail on some occasions. In 1916, at the Newcastle-on-Tyne Tribunal, an unnamed Fishbourne miner who claimed exemption as a CO, and was also a member of the Religious Society of Friends, had his appeal allowed; though at the same court an Anfield Plain miner CO's appeal was deferred until 'he worked elsewhere than in a coal mine'.[17]

The miners in the Pearce database are distributed throughout most of the coalfields of Britain. The main concentration (83%) are from the larger and more radical regions, led by south Wales (29%); and then Yorkshire, Nottinghamshire and Derbyshire (20%); Scotland (19%); and Northumberland and Durham (15%). Most (82%) of them are young men aged eighteen to thirty, the most productive of all mineworkers.

Analysis of the 'motivation' category reveals at least twenty-eight differently-stated reasons/beliefs of the miner-COs. 'Religion' often merged with pacifist and political interests. Of the clear religious groupings by far the largest were the Christadelphians (21%), usually exempted as they were willing to undertake 'work of national importance' via the Pelham or Brace schemes. This included a 38-year-old married man, C.H. Cross, directed to work underground for the Powell Duffryn Company, despite his failing eyesight, though a more humane approach was evident regarding another south Wales miner, G. Evans, who was transferred to surface work at Elliot Colliery, New Tredegar, on account of his nystagmus. Many other Christadelphians suffered the ostracism of their new or old workmates. W.J. Anstee, a Somerset pit carpenter, was approval by 'Pelham' to transfer to another pit. W.S. Taylor, from Tredegar, who had trouble in getting his 'Pelham cards' approved by his colliery, was forced to leave the pit during a strike and was directed to work on timber work or 'work in a controlled establishment' instead of a mine. Another Welsh Christadelphian, D.J. Thomas, was refused permission to work underground as his CO status was deemed 'provocative' in the context of the comb-out of miners. Among the Nonconformists groups, the largest sections related to the Methodists (19%) closely followed by the Quakers, Presbyterians and Baptists (each roughly equal in number, about 18% combined).

A 'political' group, those miners with known trade union roles, formed a significant proportion of the miner-COs, at around 15%. This cohort includes several well-known activists; Arthur Horner, George Warne, Edward Williams and Alexander Smillie, the son of Robert Smillie. Among others was a nineteen-year-old hewer from Abertillery, with a distinctive biblical Christian name, Onesimus (or 'Ness') Edwards, who went on to be elected as MP (Labour) for Caerphilly in 1939 (until 1968). A Baptist and ILP member, Edwards was also associated with the Plebs League and NCF in 1916. After allegedly going AWOL in 1916, he was arrested and handed over to the army, still refusing to obey orders. His CO history includes hard labour in Wormwood Scrubs and Dartmoor Work Centre. Ness actually escaped for a time from Dartmoor, living in the wilds of the Brecon Beacons, prior to giving himself up in 1918.[18] Arthur Brinley Roberts was another young Welsh miner CO (aged nineteen in 1916) who also progressed to become a prominent miners' and trade union (NUPE) leader. Roberts also managed to escape from prison in Northumberland but was recaptured.[19]

The trade union COs were usually treated badly by the authorities. George Minto from Spennymoor, a DMA activist, and a Primitive Methodist with an 'absolutist' attitude, refused any kind of military associations including the Brace 'opt-out' scheme. He had to endure three 'hard labour' sentences by 1919. The socialist Kelly brothers, Bartholomew ('Bart') and James, from a large West Stanley mining family, received harsh treatment for their stance. Bart's appeals were rejected, and 'referred to Brace'. 'Red Barty' and 'Stormy Petrel' as he was called by some elements of the press, was force-fed twenty-eight times whilst on hunger strike in Leeds prison. Younger brother James was no less determined to resist, refusing the Brace referral, and forced to endure two years in Hull and Durham prisons. A 'Miss Kelly', probably the Kelly brothers' sister, Jane, was secretary of the Stanley branch of the

Women's Peace Crusade during the First World War. Never forgetting his family's war resistance, James Kelly rose to become president of the Durham Miners' Association.

Twenty-three (14% of the known 'motivations') of the miners in the PRBCO database are identified as 'absolutist', most of them having to suffer brutal consequences for their principles. Somerset surface worker Fred (or Frederick) Badman (the most ironic of surnames in this context), a banksman, therefore he would have known most of the underground workmen as they descended and ascended the colliery, received *five* sentences, served over two years (with hard labour) at Exeter, Wormwood, Wandsworth and Winchester prisons. This was despite referral to Brace and the Non-Combatant Corps. A Quaker (and/or Baptist), in 1916 the 36-year-old from Old Mills, Badman had lost his brother Arthur in a roof fall at Farrington Gurney Colliery three years earlier.[20]

Wilfred Arrowsmith, aged twenty-eight, a south Wales hewer and 'Calvinistic' Methodist, (also a member of the Independent Labour Party [ILP] and NCF [No Conscription Fellowship]), was alleged to have been 'very cruelly treated in prison[s]' when detained in Cardiff, Wormwood, Wandsworth and Strangeways. He was eventually released in January 1918 'on medical grounds'. Another Welshman, Philemon Edwards, with Baptist-Quaker-NCF and ILP 'motivations' was court-martialled on five occasions. Refusing Brace, he endured spells of hard labour in Wormwood, Wandsworth, Carnarvon and Walton before being released due to illness, albeit for just two weeks, in May-June 1918. Court-martials inflicted on David Lewis, aged twenty, a Pentecostalist from the Llanelly area of Carmarthenshire, included reference to his 'disobedience' and 'desertion from duty', when serving time in Wormwood, Wandsworth and Pentonville until his release in July 1919.

All four of the Scottish miner-absolutists in the PRBCO were subjected to court-martial and prison from 1918-19. Amongst them was George Whitefield, a 24-year-old from Linlithgowshire, who was allocated to the Non-Combatant Corps, but incarcerated in Wormwood and Barlinnie (Glasgow).

For those COs with sincere religious opposition to the war, most notably the Quakers, Baptists and Methodists, some kind of non-combatant agreement was an acceptable though uneasy compromise. Often this meant service in the RAMC, perhaps as a stretcher-bearer or maybe working in a field ambulance unit or hospital; or in the NCC. Matters did not always run smoothly though. An Accrington miner, 25-year-old Thomas Whalley, had attested as early as 12 October 1914, for service in the RAMC, and was registered as a cook in military hospitals according to his military record. After almost two years he was placed on the Reserve and returned to mining but by 1918 was 'recalled to the Colours'. A Wesleyan, Whalley 'disobeyed a lawful command' and despite his CO 'classification' was court-martialled and sentenced to a year with hard labour in Wormwood Scrubs. He was subsequently placed back on the Reserve and allowed to work as a miner, at Altham (Moorfield) Colliery.

'Deserter' is a label that must have been extremely hard to bear when someone was against military service primarily because of conscience. Yorkshire miner Ernest Dinsdale, Leeds-born but working as a fifteen-year-old pony-driver in a Rotherham pit in 1911, was called to attest at the age of twenty-two, with the 3rd York & Lancaster Regiment, on 27 June 1918 when he was still a working miner. Just 5 feet 3 inches and 111lbs, he 'failed to appear' at the Barnsley recruitment office two days earlier and after arrival at Pontefract barracks refused to obey a 'lawful command'. Identified as a CO on documents among his military papers, Dinsdale was court-martialled, and sentenced to six months' hard labour in Wormwood Scrubs, but on 30 August, was released, placed on the army Reserve and sent to a 'work of national importance' under the Brace Scheme. For Dinsdale it meant the horrible prospect of forced labour at Abdon Clee stone quarry on Dartmoor. Rejecting the work, he was ordered to return to military service but deserted on 20 December, and was then sentenced to serve time in Leeds and Shrewsbury prisons. He was not discharged until 1920. Dinsdale appeared at Pontefract barracks at the same time as another CO, twenty-year-old Arthur Deakin who lived in a neighbouring village. The two may have been friends or workmates, probably at Kiveton Colliery. Deakin's service record is very similar. After court-martial and sentencing to Wormwood, Deakin 'broke the terms' of the Brace Scheme and was 'returned to the Colours' in January 1919, but was discharged the following year.

Many of the miner COs who went on to serve in the military overseas performed their non-combatant duties not only 'as required' but in many cases with extraordinary bravery. An example that recently came to prominence was the case of John Powis, featured in a special First World War edition of the BBC's *Antiques Roadshow* programme, and in presenter Paul Atterbury's associated book.[22] A hewer at his family's pit in Staffordshire, Powis was an RAMC stretcher-bearer in France and Flanders, with the 3rd North Midland Field Ambulance Division of the North Staffordshires, serving from 1915 until discharge in 1919. Initially he would have been active in the Ypres Sector and present at the start of the Somme Offensive, recovering casualties from No Man's Land on that terrible first day of July. Promoted to Sergeant in 1917, John Powis was awarded the DCM early in the same year. His citation reads as follows: 'For conspicuous gallantry and devotion to duty. He has performed consistent good work throughout, and has on many occasions shown great courage and coolness under fire.'[23] This was a most remarkable outcome for a CO, especially when commanders had been instructed to limit their recommendations for this award to 'gallantry in the face of the enemy'.[24]

The so-called 'comb-out' of miners that involved calling up for military service those who had entered colliery employment *after* August 1914, and the further call for colliery workers aged 18-32 to join the colours in the spring of 1918, resulted in another deluge of appeal applications, the Colliery

Henry Haston

The death of Derbyshire miner Henry Haston, at the age of twenty-six, on 21 September 1918, was a consequence of the harsh conditions he had to endure when sent to work in one of the most extreme landscapes in Britain, Dartmoor. Born near Alfreton, like many of his young pit mates he had joined the Colours, attesting at Chesterfield, aged twenty-two, on 11 December 1915. A much respected Sunday school teacher at his Primitive Methodist chapel, Haston was placed on the Reserve, and continued working as a miner for more than two more years until he was called for military service, to Derby barracks and the Sherwood Foresters, in the summer of 1918, as a 'combing-out' recruit. At the time, his wife Lillian was heavily pregnant with their first child, a daughter to be named Annie. As an absolutist, therefore refusing to bear arms, he was arrested, tried and sentenced to two years' (commuted to 18 months') hard labour at Wormwood Scrubs. But as a CO he opted for the Home Office Scheme, which for him meant labouring work at Princetown Work Centre, where within a few weeks he contacted pneumonia, a condition quite common there. His wife Lillian was summoned to his bedside, probably in Tavistock Hospital, just before he died. 'Henry Haston' is inscribed on a wooden plaque listing the names of seventy COs known to have died as a result of ill treatment, kept in the London office of the Peace Pledge Union; and Haston's tragic story forms part of their online 'The Men Who Said No' project.[21]

Henry Haston (1892-1918), coalminer and Sunday school teacher. menwhosaidno.org

Plaque commemorating 70 (of 81) known British COs who died through ill treatment because of their resistance to arms during the First World War. The main inscription reads: IT IS BY FAITH OF THE IDEAL THAT THE IDEAL COMES TRUE. Peace Pledge Union

Recruitment Courts (CRCs) continuing to sit until the end of the war. To complicate matters, in order to facilitate the late 'weeding-out process', which involved extraction of miners into military service according to complex quotas at each colliery – and also by ballot – itinerant medical boards were established in January 1917, linked to the CRCs. The administrative machinery groaned even further. The Northern Division mines inspector J.R.R. Wilson, in his annual report for 1918 noted that he had issued a total of 100,037 certificates of exemption from 202 sittings of the CRCs during the War. His colleague Dyer Lewis and a small clerical team of the South Wales Division had an even more massive workload, issuing 200,000 exemption cards. Lewis made the point that he authorized most of them 'in his own leisure time'.

For the miner-COs, it would be wrong to confuse a refusal to bear arms with a lack of patriotism. Invariably branded as cowards in the press, and ostracised by their former pit mates, the CO miners who were imprisoned or banished to Work Centres for their political and religious beliefs, had as much courage as if they had served on the Western Front. Their stance was especially brave in an industry where comradeship and community was an integral part of everyday life.

Recent research and publications, alongside online and interactive databases such as the Pearce UK Register of Conscientious Objectors (now with over 20,000 entries) will continue to help us to appreciate the context and reasons as to why some miners did not 'take up arms'.[25]

More than a century after the end of the Great War, we are now beginning to appreciate the amount of dissent and resistance in various forms that actually existed. Coalminers will continue to be significant voices in the unravelling story.

Chapter Seven

1917: STINKING TRENCHES AND SURREAL LANDSCAPES

To paraphrase the title of Ernest Hemingway's book, the third full year of war was like a slow tolling funeral bell that never stopped.[1] And an extract from a Siegfried Sassoon's poem encapsulates the most iconic phase of the war on the Western Front in late 1917:

> *I died in hell (They called it Passchendaele).*
> *My wound was slight,*
> *And I was hobbling back; and then a shell*
> *burst slick upon duckboards; so I fell*
> *Into the bottomless mud, and lost the light.*[2]

At home, the failure of the Somme Offensive had a devastating impact on the close-knit pit villages and coal-towns of Britain. The fact that it dragged out for almost the entire second half of a year – ending in the mud and freezing cold of November – magnified the pain and worry. The Somme was akin to the aftermath of a great pit disaster, one of prolonged grief as more and more bodies were buried at home, on foreign fields or forever lost.

Most miners and their families were not fooled by the propaganda in newspapers and magazines, on posters, and even when viewing the flickering screens in the silent picture palaces. With Lloyd George now in charge of the War Cabinet, hopes were raised and successes such as the capture of Vimy Ridge in April and Messines Ridge in June confirmed that new, well planned and executed tactical approaches worked. The Germans were kept on the defensive. However, casualties were high at Arras, one of the bloodiest battles of the Great War, with around 4,000 British soldiers lost *each day*.[3] Miner-soldiers continued to be active on both frontline and in supportive roles, including remnants of Pals battalions.

The 'war underground' continued albeit in new forms. The Battle of Messines started with the detonation of nineteen huge mines under the

German lines. Sir John Norton-Griffiths, the charismatic tunnel commander, was right when he envisaged that the landscape of the affected area would be changed beyond imagination. The exact number of German casualties will never be known, despite the oft-quoted and misleading figure of 10,000 repeated in so many sources. More realistically, the main outcome was not so much the number but the immense and immediate shock imposed, a reverberating psychological soundscape more far reaching than all the falling debris. A strategic success certainly, but Messines was also a triumph for the hard work of hundreds of miner-tunnellers over many months of preparation. Sadly, some contributors died a few days before the 'big bang', unable to see the results of their efforts. One example is that of Sapper Ernest Hull, from the colliery village of Esh Winning, County Durham, a hewer, who was fatally hit by a German shell on his way to mining duties on 3 June.[4]

And then, right at the end of July 1917, supposedly still summer, the Third Battle of Ypres began, popularly known as Passchendaele. As the cool wet late summer turned into an even wetter autumn and colder winter more and more men fell into the mud and died, to be remembered on memorials at Menin Gate, Tyne Cot and in the huge Tyne Cot Cemetery. The British total of 240,000 killed or wounded – quoted in the British Official History –

British soldiers prepare to navigate the muddy pools at Passchendaele. Taylor Library

Passchendaele, a dreadful mixture of water-filled shell craters and dead bodies. Taylor Library

was revised by the War Cabinet to 265,000, but may have been far greater. Whatever the total, the human sacrifice is unimaginable today, grief totally beyond our comprehension. However, as family historians will testify, stories about loved ones passed down the generations.

Passchendaele completely changed our perceptions of the making of a battlefield landscape, with its eerie moonscape of craters and shell holes filled with a dreadful mixture of foul water, mud, blood and bodies. Thus the year 1917 and 'Passion-dale' with its horrendous casualty figures and somewhat inconclusive outcome has become *the* symbol of the Great War, recognised in two new, modern ultra realistic films: Sam Mendes's *1917* (2019) and Edward Berger's German version of *All Quiet on the Western Front* (2022).

The stinking trenches and surreal landscapes of blood and death remain powerful images long after leaving the cinema or viewing at home. Much has been written too, about Passchendaele, from the classic mid 1950s writings of Leon Wolff to numerous modern reappraisals of the conflict, including Richard Van Emden's compelling prose based on the original words and photographs of those who were there.[5]

Towards the year end, on 20 November, the Battle of Cambrai began, with 378 British Mark IV tanks leading the advance towards the Hindenburg Line, the first orchestrated, massed use of tanks in warfare. The Line was well

The attack at Cambrai was to prove that the tank was an essential weapon with the capability to breach the barbed wire defences and bridge trenches.

breached but by midday almost half of the tanks had succumbed, largely victims of mechanical failure, and by early December much of the British territorial gains were lost to German counter-attacks, both sides suffering an estimated 45,000 casualties.[6]

The Welsh mining entrepreneur, Lord Rhondda (David Alfred Thomas) was appointed by Lloyd George as Food Controller in the summer of 1917, and via DoRA, he went on to introduce formal rather than advised rationing in 1918.

The prolonged winter of 1917-18 exacerbated poverty in the coalfields.[10] Earlier, in late 1916, in the small south Yorkshire coal town of Mexborough, miners' union representatives and local families met in the Empire cinema to protest about the cost of living.[11] Here, at least seventy-three local soldiers had been killed in action or died of wounds during the Somme campaign, mostly miners. When the wage earners had gone there was little support for grieving widows and deprived children. For many, the close bond that always existed in pit villages was the only consoling feature of another terrible year ahead.

Everywhere in the coalfields there were signs of mourning, especially close to the bigger pits where hundreds of miners had rushed to volunteer. Many of the bereaved had little consolation regarding 'missing' and 'no known grave' reports, confirmations of deaths often received many months afterwards.

Parson pitmen

During 1917, occasional reports appeared in the press about parsons, who were normally exempt from conscription 'responding to [appeals for] volunteers for National Service' by working in the mines, presumably to help maintain coal supplies. The Reverend Ernest Edward Johnson, priest-in-charge at St Luke's in New Rossington, Doncaster, began working on a three-shift rota at Rossington Main Colliery. He was given an arduous job underground, as a 'filler', shovelling coal into tubs or wagons and had to walk almost a mile underground to get to his 'stall' or place of work. What's more, he was under the supervision of his own church choirmaster! Eventually Johnson got a commission, as a Captain in the Lancashire Fusiliers. But tragedy struck. Serving as a chaplain, he succumbed to pneumonia in France, on 1 December 1918 and was buried in Le Cateau Communal Cemetery. He was aged thirty-three.[7]

Reverend E.E. Johnson. Doncaster Chronicle

Askern Colliery, a large pit, also in the Doncaster area, had a senior churchman, Canon Thomas Rolfe, vicar at Kirk Bramwith and Barnby Dun, rolling up his sleeves to work 'on the screens', picking debris from the coal. This was a notoriously dirty, noisy job, often associated with a fair bit of cursing from the man-in-charge! Two other intrepid local parsons, William Rutter and Montague Allen, like Ernest Johnson cited above, filled coal underground.[8]

In Wales, at least three Nonconformist ministers were reported to have 'gone to work in coal mines' and another source noted that 'some ministers are engaged in surface work in coal mines and about collieries'.[9] One wonders how widespread and how long the use of 'pit parsons' was!

Three 'vicar-miners': (L-R) Rolfe, Rutter and Allen. Doncaster Gazette

Stinking Trenches and Surreal Landscapes • **157**

Typically, Catherine, the widow of John Glynn, a former Barrow Colliery and Barnsley Pal, who was killed on the Somme on 1 July 1916, pleaded via the local press for official news of her husband. Glynn's possessions were not returned to her until a year later. What a time she must have had looking after six children; and how on earth did she manage to cope afterwards.[12] Further north it was a similar story. Corporal Ernest Chance, serving with the Tyneside Scottish, a Brandon Colliery miner, and son of a County Durham Methodist minister, lost his life on 1 July 1916 but his widow and children had to wait almost ten months to get confirmation of his death.[13] Two new insightful accounts help us to appreciate all the heartaches and the long-term problems that were faced by the war widows: Andrea Hetherington's *British Widows* and Richard Van Emden's *Missing*.[14]

Barnsley, a coal town in mourning

Barnsley people, in the wake of raising two Pals battalions and its plethora of volunteers to all of the services, suffered badly. Sampling the months in the weekly editions of the *Barnsley Chronicle* – front pages mainly – the number of reported fatalities increased alarmingly in the spring months. In April, almost one hundred 'local heroes' (killed in action) were named, a good number of their family's demises described in some detail. But during May the toll doubled, a reflection of the bloody Arras campaign. Between 70-85% of casualty reports (where occupations are mentioned) were former miners. Extra columns and smaller print were requested by journalists and approved by the editors in order to keep up with news from the fronts. A steady flow of typesetting transitioned into a torrent, a rush to meet copy deadlines worse than ever before. More black ink, more black fingers, in coal-black Barnsley. Families raced to buy, borrow and share the town newspaper that had become *the* fountain of information, even hope. News was devoured and then read again and again.

Most reported fatalities were described as 'killed in action', the news often 'softened' by the usual 'no pain' phrases about the individual from a senior or commanding officer. But there were also a significant number of previous 'died from wounds', 'presumed dead' and 'missing' reports that were now confirmed as deceased. Even for a town used to dealing with major and day-to-day tragedy of mine-related deaths, the impact was both devastating and prolonged. People were numbed, hardly able to speak for weeks on end. Hard-up families scraped a few shillings and pence together in order to insert carefully composed tributes to loved ones in obituary columns that in some weeks extended to three full columns. Tears of sorrow fell on to pencilled drafts written on bits of damp, crinkled paper. Although, as mentioned above, the significant increase in the number of reported casualties reflected news about the Battle of Arras, and news about Somme casualties continued. It was cruel, late sorrow for forever grieving families.

During the last few months of the year reports of war fatalities steadied to about sixty a month, peaking in November when at least a hundred deaths are mentioned, largely a reflection of the growing impact of the Battle of Third Ypres (Passchendaele).

Double grief: Private Joseph Bell and Sapper Thomas Frederick Bell

What a horrible Christmas it was for Matthew Bell, a 46-year-old hewer and his wife, Lily, residing in central Barnsley. The commanding officer of their youngest son, Joseph, a Private in the 2/8th (Territorial) Battalion of the Sherwood Foresters (Nottinghamshire and Derbyshire Regiment) had informed them that their son was missing, as from 26 September 1917.[15] There's no subsequent mention of Bell in the columns of the *Barnsley Chronicle*, so it may have been more than a year before he was officially declared as deceased. Joseph, probably 'Joe' to his family and friends, is listed on the Tyne Cot Memorial to the Missing, one of thousands who died in the Passchendaele battle without any known grave. Enthused no doubt by the example of his older brother, Thomas Frederick (see below), Joe enlisted well underage when he was a sixteen-year-old miner at Monk Bretton Colliery.

Joseph Bell. *Barnsley Chronicle*

Sapper Thomas Frederick Bell lost his life two Christmases earlier, just weeks after his younger brother had arrived in Flanders. Initially, 'Tom' was probably a Territorial, in the 5th/York and Lancaster Regiment (and not 13th Barnsley as stated in De Ruvigny's Roll), mustering in Barnsley in September 1914, a few weeks after his marriage to Fanny Dixon in St Mary's church. In Flanders from the spring of 1915, he was soon transferred to the Royal Engineers, and attached to the 172TC. He would have seen action in the Bluff/St Eloi/Hill 60 areas near Ypres. Tom's date of death – 19 December 1915 – is a significant one as it coincides with the first use of phosgene gas by the Germans nearby, at Wieltje, a mile and a half north east of the future site of the Menin Gate Memorial which bears Bell's name.[16]

Thomas F. Bell. *Barnsley Chronicle*

'One of our best pals': Private Harry Glover

Sixth child in a large mining family from Stairfoot, Barnsley, Harry worked as a rope repairer in 1911, probably at Grimethorpe Colliery. He enlisted in the KOYLI's at Wath-upon-Dearne on 7 September 1914 but got transferred

Stinking Trenches and Surreal Landscapes

to what was probably his preferred battalion, the 13th York and Lancasters (Barnsley Pals), and was in France by early March 1916. He appears to have been affected quite badly by bouts of scabies, a common occurrence in and around the trenches but – to the immense relief of his family – Harry survived the Somme Offensive. Unfortunately, a few months later, he lost his life in fighting around Gudgeon Trench, Pozieres on 11 March, aged twenty-four. A published letter 'from a comrade' to his mother, Mary Glover, explains his demise and includes a now-familiar last sentence rider:

Harry Glover. *Barnsley Chronicle*

> He was one of our best pals, and we are sorry to lose him. He was struck by a piece of shrapnel whilst doing his duty near the German lines. We are pleased to say that he did not suffer any pain, his death was instantaneous, and you have the consolation in knowing that he died for King and Country.[17]

The strength of the 13th Barnsley was then thirty-three officers and 940 other ranks. According to the War Diary, Glover, in B Company, was the '1 o. [other] rank killed' (four others injured). The previous day, 'Captain Normansill [*sic*] died of wounds'. 'Jack' (John) Normansell was well known in Barnsley, as he was a grandson of the late John Normansell, one of the founders of the Yorkshire Miners' Association, serving as general secretary, a miners' leader of national importance. Prior to his commission, Normansell was completing a mine manager's course in Sheffield.

A selfless stretcher-bearer: Private Claud Leatham

Leatham had the great misfortune of being one of the earliest Barnsley casualties at the Battle of Arras, losing his life on the first day of action, Easter Monday 9 April 1917. His 'death in the snow' was typically unselfish for those serving in the ambulance corps, hit by artillery fire whilst carrying a wounded officer through No Man's Land.[18] Single, aged twenty-five, Claud and his family were resident at Great Houghton, in Darfield parish prior to enlistment. Like his father George Leatham, he was a hewer at Houghton Main Colliery. Interred in Cabaret-Rouge British Cemetery, Souchez, France, his family managed to afford tributes inscribed on gravestones in churchyards at Darton and Great Houghton. In addition, and more immediately, there was a 'paid-

Claud Leatham.
Barnsley Chronicle

for' memorial tribute inserted in the *Barnsley Chronicle*. This, at a time when his last resting place was unknown.

Unlucky young leader: Acting Corporal Francis ('Frank') M. Mullins

Frank Mullins.
Barnsley Chronicle

Mullins, of an Irish heritage, from a large mining family living at Wombwell, worked at Houghton Main, one of the most successful collieries for military recruitment in the Barnsley coalfield. Joining the battalion in September 1914, he would have survived action on the Somme during 1916 and the following April was active at Arras, when, as an Acting Corporal he must have impressed his officers. During the Third Battle of Scarpe he was placed in charge of a small group of Pals executing bombing raids.[19] Frank's cohort had captured a trench and began securing their position when on 11 May he was unlucky enough to be the target of a fatal bullet which penetrated his helmet.[20] His body was never recovered, but he is remembered on the Arras Memorial and on local churchyard family memorials.

Reliable, persistent and popular: Company Sergeant Frank Ashton

Frank Ashton.
Barnsley Chronicle

From Wombwell, Barnsley, Frank was an underground 'signal boy' according to the 1911 census, his father also given a specific mining occupation by the enumerator: a 'dayman shot-lighter'. By the time of his enlistment at the age of nineteen on 21 September 1914, young Frank, at almost 5 feet 10 inches, tall for a miner, had probably advanced to coalface work, maybe at Cortonwood. On the Western Front from 16 November 1915, he progressed through the ranks, serving as a Sergeant from 15 July 1916. A few days later he was wounded, a gunshot to his right shoulder – and yet again on 11 August (shoulder and head) – but after a second recovery phase advanced further, to that of Company Sergeant Major, in January 1917. Wounded in the shoulder and hand again on 20 May, Frank died nine there days later, at No. 20 Casualty Station and was buried in the Sunken Road military cemetery at Boisleaux-au-Mont, France. A glowing testimony from his CO was perhaps some consolation to his grieving parents, John and Lucy Ashton back in Wombwell, and included the following:

> . . . he was in splendid spirits as ever, and kept the men singing and cheerful . . . and getting [them] back into position until we moved forward to the assault . . . nobody could have found a truer or more

reliable man – always popular, a splendid example of a soldier in action, and a man to be proud of and emulate.²¹

Messenger runner to 'a swearing parson': Private John Howarth

John Howarth.
Barnsley Chronicle

George De Ville Smith.
Barnsley Chronicle

The Howarths were neighbours of my maternal grandparents, in the large pit village of Royston. John Howarth followed his father into underground work at the large Monckton colliery complex. One of the early Barnsley Pals volunteers (No.457), when on the Western Front John became a 'messenger boy' for one of the great characters of the first Barnsleys: Captain Reverend George De Ville Smith. This 'swearing parson' from St Thomas's, Worsbrough, commanded D Company, and was known by his soldiers as 'Devil Smith' on account of his propensity to curse at every opportunity. Howarth was wounded in the July 'push' on the Somme but unlike De Ville survived the 1st of July, only to be killed in action on 2 November 1917.²² The War Diary of the 13th for 2 November contains a brief 'anonymous' entry to '1. O.R. [other rank] killed' that must refer to Howarth, when in the Archeville Sector trenchers. News reached his parents shortly before a 'Local Heroes of the War', featuring four soldier images, appeared on the front page of the *Barnsley Chronicle*.²³ Unmarried, John Howarth, aged twenty-two, was buried in Roclincourt Military Cemetery, France.

***

Despondency and dread permeated mining communities as the year progressed. Unofficial strikes, unfair production demands from the War Office and the coal owners stirred unrest. Bereaved families had little prospects as fathers, husbands brothers and sons returned from leave or discharge from military service, carrying a plethora of mental and physical scars. Some managed to return to their pits, though were not always able to resume their previous job. Less pay for homecoming heroes. Others, about an equal number, for a variety of reasons, could not even return to mining. Youngsters and alleged shirkers from a numerous trades became 'pop-up' miners, regarded by the unions as conscript dodgers and unsafe, threatening the safety of others.

Conflicting demands: 'Coal for war' v 'miners for war'

In anticipation of the heavy fighting ahead, as early as January 1917 the War Cabinet acknowledged that the army was short of 100,000 men. Although a quarter of a million miners remained in military service it was decided to call-up 20,000 more as part of its wider strategy, starting with those deemed as unskilled and the new entrants. By August, the owners and miners' unions were instructed to release 21,000 men under the new scheme, but now as part of an even greater requirement: there was a 40,000 (Class A) quota. This was an extraordinary policy turnaround. By then, 19,000 had already been 'liberated' from the pits but their selection was a bureaucratic nightmare to operate.[24] Single men, aged between eighteen and twenty-five were targeted, ironically the fittest and most able 'coal-getters'. Exclusions included managers or 'officials', underground hauliers, fitters, mechanics, blacksmiths and winding engine operatives, presumably so as to ensure that mines could continue to operate as normally as possible. Each mining district or county was allocated a quota calculated on the basis of the total number of persons employed. In 'troublesome' south Wales 500 miners were required for the military from the Amman Valley, around twenty-five from each pit. Imagine the in-house discussion that followed, each colliery having to decide 'who to go' or who to stay.[25] In Scotland, enlistments in Lanarkshire were well short of a set target of 1,500, the unions unable to agree on selection criteria.[26]

Numerous examples arose of decisions that left colliery families bereft of their main income provider. In Hexham, Northumberland, a 'banksman' (an operator of cage loading gear and shaft signals, an indispensable job) was excluded exemption even though he was the main support for his disabled parents.[27] Common sense reigned in extreme cases, thank goodness. A handicapped Ludworth, County Durham, miner, unable to walk more than five yards, and off work since 1913, was given absolute exemption. In many cases the decisions were at best harsh and at worst cruel, situations hard to understand when judges included men with local knowledge. Generally, the fine 'combing-out' and production-demands lines were tested to breaking point, the state and the army pulling against each other as much as each dare.

A compromise of sorts had to be agreed. The two miners' federations (for England and Scotland, and for Wales) were concerned about the twin demands of 'coal for the war' and 'miners for the military'. In September, some 160 delegates attending a conference at Blackpool recommended that it would be men who had come into the mines from other industries – since 4 August 1914 – who should be extracted first, irrespective of their job roles. An amendment from the south Wales miners stating that the unions should not assist the War Office in the selection of men was heavily defeated. The final, passed resolution was for all Class A workers employed in and about the mines between the ages of eighteen and twenty-five, with no exemptions, should be applied, but only to come into effect 'if sufficient men are not obtained under the terms of the resolution'.

The combing-out process was bedevilled with complexities and inconsistent interpretations. Beleaguered Colliery Recruiting Courts received huge numbers of exemption applications, too many to deal with fairly. Under the heading 'A Miners' War' (or similar), the comments in one coal town newspaper were repeated in similar words in most other coalfields:

> Sometimes when a special job was required at short notice, and it would have produced too much dislocation to collect men from the ranks, a telegram to this country has led to a contingent being sent straight to Flanders from the coal-pits at home, and men over sixty years have been found serving their country with the pick not far from where their own sons were fighting for it with bayonet, ALTERING THE WAR MAP.[28]

The same writer extolled the success of the 'blowing up of the Messines Ridge' as a successful example of the miners' contribution to the war. Therefore, *in addition* to the 40,000 quota, miners – even veterans (providing they had special skills) could be fast-tracked to the Front.

By the end of the year the Home Secretary acted on the federations' resolutions, scrapping exemption certificates already issued to 'pit top workers' in favour of all men working in and about the mines employed after 4 August 1914. The frustration and disappointment of those who had gone home from the courts to their families thinking that they were free to continue to work as miners must have been enormous. Thus, despite the Military Service Acts and its amendments, thousands of miners were pulled, indeed fast-tracked to the Western Front 'on demand' if their colliery or area received a telegram request from the War Office.

The government and the mines inspectorate continued to demand that coal production should be maintained and increased 'for the war effort'. All manner and means were deployed in order to 'save coal' and bolster production. At its most cruel were ad hoc comments in the press, including numerous printed in the conservative coal industry journal the *Colliery Guardian*, stating that miners should 'work harder', and suggesting fines or dismissal for having a day or more off work. Most of the older men were weary if not exhausted. There was also a plea – following the death of a miner in an accident – that workmates should not take a day off 'in respect', a long tradition in some coalfields, including the huge Yorkshire region. The YMA, recognising the importance of maintaining coal supplies for the army and the home, supported the request to suspend or stop the old practice. Thus, at Hemsworth, in March, the *Colliery Guardian* reported that the miners at local pits had 'ensured' 1,500 tons of coal 'for munitions' by continuing to work after a pitman had been killed falling down a shaft.[29] In recompense, the owners presented a cheque for £25 to the deceased's widow.

The miners and their unions did what they could in support of the war and also help a burgeoning number of bereaved families. The unions transitioned into social services. The YMA paid out benefits to the families of 2,256 of its members killed in the war.[30] By autumn, financial help became more urgent, as 3,115 of union members had now been 'killed in the war'. Yorkshire miners themselves handed over to its association their bonus money for working during a holiday period, a considerable sum of £17,500, to be used for 'war charities'. This was in addition to the provision of fifty motor ambulances for the Red Cross at a total cost of £35,000. So much for the 'unpatriotic miners' criticism.

Thus miners, seemingly ring-fenced against conscription, from the spring of 1917 continued to be called to arms if they had specialist tunnelling, medical or rescue skills that would benefit their service on the Western Front. Despite the great loss of miners 'to the military', employee numbers in the coal industry began to stabilize, indeed even increase. The mines inspectorates' annual report for the year 1917 stated that 1,041,840 persons were employed 'in and about mines', 823,936 of them working underground. The latter represented an annual increase of 19,217 (16%). Interestingly, in 1917, 11,297 females worked on the pit tops, and a similar increase in numbers was reported: 1,350 new recruits (15%) in a year.[31]

The buoyancy in the number of people employed in coalmining was a consequence of at least one in five miners being rapid replacements for those who had volunteered. Thus pits had a large number of new entrants, from school leavers and factory hands to retail workers and ex-teachers; and of course those formerly unemployed. Sceptics and miners' union leaders warned about the implications of so many 'green incomers' on safety. Subsequent statistics tend to support their concern. In 1917, 1,370 persons lost their lives in coalmines (1,214 in underground accidents), an increase of fifty-seven from the preceding year. This was a horrendous toll, South Wales and the Yorkshire and North Midlands regions – with a shared total of 800 fatalities – were hit hard.[32]

But, despite of all the dangers of working down a mine, buzzing in the brains of many union officials, politicians and experienced mineworkers was an abiding question: had the 'newbies' got jobs in mines in order to escape conscription?

Messines and the 171 Tunnelling Company

171 Tunnelling Company (171TC) of the Royal Engineers – many of them Welsh miners – continued to be at the forefront of mining operations during preparations for the decisive battle of Messines Ridge. The climax was the great detonations of 7 June 1917 when nineteen mines packed with almost a million pounds of explosives were successfully detonated at 3.10am. The main impact was the immense shock that the event generated, most German casualties accruing from British artillery assaults.[33] Post-Messines, the 171TC were involved in 'non-contact' mining, supporting the infantry in new-won ground, but this was far from being an 'easy' lull in action as casualties

Caterpillar Crater, near Hill 60. Brian Elliott

continued to be sustained. One of them, Sapper Henry Thomas Parfitt (see outbox), whose demise was featured in a local exhibition about the Great War – and also in Ritchie's Wood's *Miners at War 1914-1918* – is a typically sad example. Unfortunately, as Ritchie explains, the exact circumstances and location of Parfitt's death, on 16 August 1917 (like several others about this time) cannot be verified, as the weekly war reports were absent for over three months.[34]

From offensive mining at Hill 60 in 1915 and the undermining of Messines Ridge in 1916-17, to three years later, when they were involved in a non-contact mode in supporting the British infantry's advance and the German Offensive, 171TC's contribution was considerable. But their role is perhaps best appreciated through the stories of individual soldiers, and this is why the meticulous research of Ritchie Wood is so relevant to our proper understanding of their importance. Wood's tabled analysis tracks no less than twenty-eight 171TC Welsh tunnellers' 'experiences' on the Western Front, from 1915-19.[36] Almost half were classed as 'KiA' (killed in action) or DoW (died of wounds), the remainder returned to their units and in due course discharged, including several who had been gassed or suffered from acquired nystagmus (eye ailment). As Ritchie says in his conclusion, one wonders about the extent of feelings that existed in the minds of those few who were able to return to mining, especially when in conversation with new and former workmates who had not enlisted.[37] Another unknown was if they were able to return to their previous job, hewing for example being a role that demanded a high level of strength and fitness. Lighter work meant far less pay, hardly compensated for by a modest army pension.

Sapper Henry Thomas Parfitt

From Orford in rural south Gloucestershire, Parfitt moved to Monmouthshire, for work at Elliot Colliery, New Tredagar. He volunteered at Newport, enlisting in the 8th Battalion, South Wales Borderers, probably in the late summer or autumn of 1914. Embarking to France on 25 February 1915, Parfitt was soon transferred 'in the field' to the 171TC. Due to his mining background, he would have been a good man to deploy in the preparations for the 'great blow' of 7 June and its aftermath. A further advantage was that he was experienced in the use of mine rescue apparatus, having gained two proficiency certificates. It appears that he suffered a wound and shell shock whilst with the 171TC, his death occurring on 16 August 1917, in 22 Field Ambulance. An accolade came to his family when he was awarded the Military Medal, posthumously, 'for heroic work in connection with the 'great explosion of the German position at Messines', though it was not gazetted until 13 March 1918.³⁵ Sapper H.T. Parfitt MM was buried in the Menin Road South Military Cemetery, Ypres. The inscription at the base of his memorial reads: THEIR GLORY SHALL NOT BE BLOTTED OUT.

Sapper Parfitt was twenty-seven years old when he lost his life serving with 171TC on 16 August 1917. Brian Elliott

The Royal Engineers Grave at Railway Wood³⁸

The distinctive CWGC Royal Engineers' 'Cross of Sacrifice' grave memorial on the Bellewaerde Ridge near Zillebeke, pays tribute to eight Royal Engineers of the 177TC and four attached infantrymen killed in the intense underground fighting during the defence of Ypres, their bodies never recovered. Seven of the men were lost in 1917, six of them former miners or from mining families.

The first was the tragic case of a young north Wales miner, barely out of his teens who, after enlisting with the 17/Welch Fusiliers was attached to an associated Pioneer Mining Company and then to the 177TC on the Western Front. **Private/Sapper Thomas Edward Davies** whose family came from Overton-on-Dee, Wrexham. Standing just 5 feet 5 inches and weighing 8 stone 6 pounds, he lost his life in unexpected circumstances on 25 February at 5.10am when a German mine exploded close to number 6 workings where he was placed. The blast formed a substantial surface crater on No Man's Land about 100 feet in diameter and 30 feet deep. It is thought that German 'listeners' had become aware of a lull in the remedial operations that had been taking place, triggering a race to blow a defensive mine.³⁹

Morley (Leeds)-born **Corporal/Sapper Sidney Firth** was an eighteen-year-old pony driver in 1911, probably at the same pit as his father, an experienced

Railway Wood Grave, with an unmistakable mine crater in the foreground. Brian Elliott

hewer, the family residing at Stanton Hill, Ashfield, Nottinghamshire. This area attracted many miner migrants from Yorkshire, Derbyshire and especially Staffordshire, to relatively new and developing pits, such as Sutton and Silverhill. The colliery owners were the well-known Stanton Iron Company who built company houses for the incomers. Sidney appears to have been a direct recruit into the Royal Engineers, and in particular the 177TC, in France and Flanders from November 1915. He was killed whilst on listening duty at the end of gallery 11, when the Germans blew a 'revenge mine' at 6.45pm on 9 March 1917. His closeness to the enemy position almost certainly meant an instant death, perhaps a small consolation to his family. Two other men were fortuitously found 'alive and well' after a 30-hour entombment but Sidney's remains lay undiscovered.[40]

Sapper Firth's name on Railway Wood memorial.
Brian Elliott

Two more Welsh fatalities occurred on 9 April 1917, in similar circumstances. They were **Corporal/Sapper Daniel Brookes Evans**, a twenty-year-old mining engineer from Dowlais, south Wales, and **Private Richard Roberts**, from Bethesda, Bangor, north Wales, probably a quarryman. The pair had joined the 16/Royal Welch Fusiliers (RWF) and disembarked onto French soil slightly after Sidney Firth's arrival, in early December 1915. Surviving the Mametz Wood attacks in which two out of every three of the RWF were casualties, their fortunes changed quite dramatically, when both – now attached to the 177TC – were killed by a camouflet fired by the Germans at 12.50pm on 9 April 1917. The gallery in which they were positioned was damaged so badly that it was beyond repair and unable to be accessed. Three other men were wounded. The wet conditions were thought, according to the Mine Explosion Report, to have been the reason why the Germans had not been heard.[41]

Sappers Evans' and Roberts' names on the Railway Wood memorial. Brian Elliott

Originally from a large north Derbyshire mining family, records relating to **Sapper John Henry Cotterill's** background are sparse but he was attached to the 177TC, probably after service with Nottinghamshire and Derbyshire Regiment ('Sherwood Foresters'). One of the more mature ex-miners, he was thirty-seven years old when he lost his life on 22 July 1917. The tunnel in which he was working collapsed, burying him, after the ground above was hit be a German 'high trajectory' bomb, according to a 'amateur historian' source.[42]

The last tunneller to be remembered on the Railway Wood memorial for the year 1917 was **Corporal/Sapper George Arthur Woolley**. In 1911, aged thirty-four, he worked as a 'roadman' in a coal mine and was resident at Selston in the Ashfield district of Nottinghamshire, along with his wife Martha and their five children. A dugout collapsed on the corporal following an artillery hit just at a time when – in typical miner-mode – he was attempting to rescue a comrade, his adjutant explaining in his 'regretful' letter to his distressed wife that the company had been unable to recover his body for burial. Martha had to cope on her own with six small children.[43]

Tunnellers, gas testing and mines rescue

The danger of gas in tunnelling was well known and painfully evident during offensive and defensive operations. In one six-week period, one tunnelling company incurred 150 casualties – including sixteen fatalities – due to carbon monoxide exposure. Another, in just one month, had twelve of its tunnellers killed by gas inhalation, twenty-eight others hospitalised and sixty more suffering from the minor effects of gas poisoning, though the latter remained in active service.⁴⁴

After the commissioning of Captain D. Dale Logan to the RAMC, 'gas safety and awareness' began to be taken more seriously, at least from an organisational point of view. Logan was advisor to GHQ on matters relating to the health of the miner-recruits to the new tunnelling companies of the Royal Engineers. His formidable remit included the establishment of an organised system of rescue work and gas protection on the Western Front. Improvisation, such as the adaptation of face coverings from captured German masks gave way to more efficient 'mouth filtering' appliances.⁴⁵

However, by 1917 the tactic of using shallow offensive-defensive mining had waned, replaced by the sinking of deep shafts, timbered part-lit tunnels and 'dugouts' for troop accommodation. This change meant the latest, most reliable kind of mining technology needed to be available and deployed, especially for search and rescue.

As improvements continued, the tunnellers and the rescue men, often referred to as 'protomen' because of their 'Proto'-branded breathing gear, had at their disposal just about all the equipment that was available at the coalfield-based mine rescue stations. It was a mix of old and new technology. Images of colliery rescue teams produced in postcards about this time often show canary cages and 'flame lamps' alongside the new battery-powered electric lamps.

Testing for monoxide gas was of immense importance for safety in extreme war conditions. Colourless, odourless and tasteless, it was *the* silent and deadly death hazard that could kill dozens of men after an explosion. Animals with sensitive physiology, such as canaries and mice, were deployed in urgent situations during the earlier months of tunnelling.⁴⁶ For the miners, the canary was the most trusted sensor of gas recognition. In the chaos and urgency of war, canaries were rushed down shafts and tunnels in small cages and would, hopefully – sometimes miraculously – revive later, in fresh air. Each mine rescue centre on the Western Front was responsible for keeping its own supply of 'yellow birds', looked after for by an experienced carer, usually a former miner. In some cases the claws of the birds were routinely clipped short, increasing their sensitivity to fall off perches

Proto breathing apparatus. Brian Elliott

in gaseous situations, an extraordinary example of attention to detail by the handlers.[47]

Principally for the aid and recovery of trapped miner-tunnellers, 'suits' of Self-Contained Breathing Apparatus' (SCBAs) were made available via

Dr J.S. Haldane (1860-1936): clinical physiologist

John Scott Haldane, a physiologist of outstanding ability, was called to the Western Front after the Germans used poisonous gas for the first time against Allied troops in the Ypres Sector on 22 April 1915. Quite quickly, Haldane was able to identify the German's use of chlorine and phosphene gas and within a few weeks had designed, and tested a basic 'anti-gas mask'. Known as the Black Veil Respirator (BVR), it was to replace the impromptu muslin-wrapped cotton wool mouth protector that had been hurriedly issued to troops on 1 May. Haldane knew his innovation was a temporary one, more a first response rather than a complete solution. Cluny MacPherson's 'smoke hood' and then a variety of what became known as 'box respirators' superseded the BVR, but Haldane's innovation was the start of a journey of continuous improvement to combat the growing menace of chemical warfare.

J.S. Haldane (detail from Martin Goodman's *Suffer and Survive* book) and Haldane testing his own equipment in a coal mine. Martin Goodman

Haldane was greatly respected in the mining industry because of his experiments and discoveries concerning miners' exposure to deadly 'afterdamp' in the wake of underground explosions. His scientific conclusions were based on actual attendance at several Welsh pit disasters, carrying out blood tests on survivors. Most remarkably, he also subjected himself and his son to inhaling carbon dioxide (CO^2) in order to develop a better understanding of its physiological impact. He was therefore well aware that inhalation of this toxic gas was the main cause of the death of miners after 'disaster explosions'. In a war situation matters were even more hazardous and complex. The Germans started to introduced poisonous gas into known British workings, the vapour seeping into layers of chalk, creating a potentially deadly atmosphere during tunnelling or at listening posts.

Haldane was the first professional medical person to recommended the use of sensitive animals such as mice, but especially canaries, as detection tools for gas. His solution to the deployment of the birds and their reuse was a most ingenious invention, a small portable 'canary box' equipped with a mini oxygen cylinder, enabling the bird to be revived. The Haldane Canary Resuscitator, became one of Siebe Gorman's great successes for usage after explosions and fires in mines and warfare. They are often evident in images of rescue teams and their equipment.[48]

The Haldane canary resuscitation cage.
Brian Elliott

Showered with a variety of prestigious professional awards for his work with miners, tunnellers and divers, Haldane was elected president of the Institution of Mining Engineers and headed mine research labs in Doncaster and a noted centre of mining excellence: Birmingham. The city university bestowed him with an honorary professorship. A front-line scientist of the war, John Scott Haldane remains a somewhat forgotten hero of the Great War and deserves to be mentioned more often for his extraordinary contributions.[49]

strategically-placed rescue stations where two men were on duty at all times. The aim, in active mining sectors, was that no shaft was allowed to be sunk further than 200 yards from a station. The stations themselves stocked an array of equipment, very similar to those used at the new mine rescue bases in the home coalfields. Experienced instructors and 'brigade men' were commissioned into the Royal Engineers from their home stations, so as to ensure that the users were trained for the operation of breathing and associated appliances, and also had First Aid experience.

This sophisticated arrangement would not have materialised but for the recruitment and expertise of trailblazing mine rescue professionals. Arthur B. Clifford (see outbox) was the most outstanding example, achieving amazing success. This, despite being given a nigh impossible workload, initially. From May-October 1915, he was the sole instructor, with a remit to train 3,000 men every month. He was supplied with just thirty sets of Proto gear. More former superintendents from the rescue stations were fast-tracked into army service accordingly, and SBCAs rushed out in carefully-packed wooden boxes. Lieutenant Rex Smart, a former instructor at the Dudley Mine Rescue Station, was one of a subsequent cohort of newer recruits.[50]

When underground warfare started there were several brands of SCBAs available (Meco, Draeger, Weg, Aerophor, Salvas) but it was Proto that became the default appliance for deployment on the Western Front. Its emergence began in about 1907, when a remarkable young engineer, Robert Davis (later Sir Robert, 1870-1965), an inventor and diving equipment pioneer, greatly improved a version of breathing apparatus created by the Frenchman Henry Fleuss, re-naming it 'Proto'. Manufactured and marketed by Davis' London company, Siebe Gorman, it was reliable and allowed about 45 minutes per person usage.[51] Chest-mounted, with twin oxygen cylinders, and weighing up to 35lbs, the Proto kits also included accessories such as flexible oxygen supply tubes, nose clip, mouth piece, goggles and a skull cap.

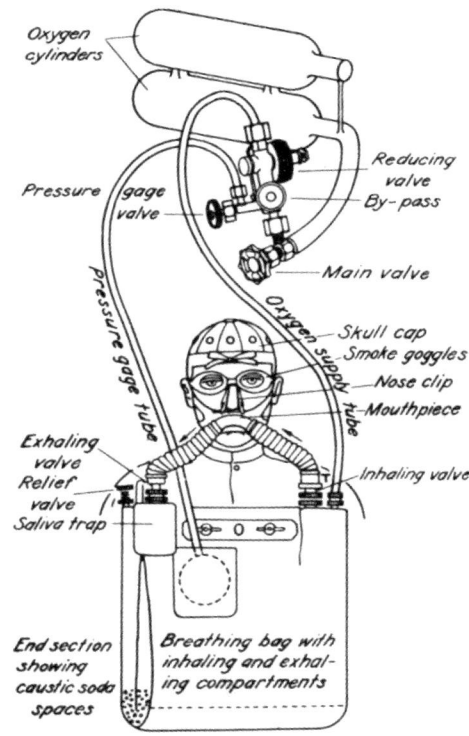

Diagram of Proto SCB apparatus.
Bulman and Mills

Arthur Bernard Clifford (1894-1961): safety in mines pioneer[52]

Born in the Barnsley pit village of Dodworth, Arthur Bernard Clifford (1894-1961) followed in the footsteps of his father Walter Clifford (1862-1923), a miner with a Nottingham heritage, who was the second person placed in charge of Britain's first mines rescue station, located at Birdwell, near Barnsley, generally known as the Tankersley Mines Rescue Station.[53] In 1910, Walter had succeeded the first professional instructor, W.T. Winborn (variations of his surname in newspapers) who had been appointed as superintendent at the new Crumlin Rescue Station, Monmouthshire.[54] Clifford senior had been awarded the prestigious Edward Medal (First Class) aka 'miners' VC' for his rescue work at the Hampstead Colliery fire in 1908.[55]

Arthur Bernard Clifford (1894-1961).
Philip Clifford/heroesofmine

Stinking Trenches and Surreal Landscapes • 173

Arthur Clifford is standing on the left of this military photograph from c.1915-16.
Philip Clifford/heroesofmine

The Cliffords moved to Staffordshire only a year or so after Walter's Tankersley appointment, as he was asked by the North Staffordshire Coal Owners to establish a rescue station at Stoke-on-Trent. Arthur Bernard Clifford, only eighteen, was appointed Assistant Instructor to his father, and quickly became familiar with Proto breathing apparatus and a variety of other safety equipment used in mine rescue.

In 1915, Arthur's reputation was such that he was targeted via an urgent telegram to go to the Western Front, where the proper use of breathing gear was thought to be vital tool for rescue and recovery operations in underground warfare. Fast-tracked in a couple of days, and given the rank of Lance Corporal in the Royal Engineers, 'AB's' initial duty was to demonstrate and train thousands of miner volunteers in the use of Proto gear, initially at a makeshift depot a couple of miles or so from the Front.

More permanent army mines rescue stations or schools were then established by Clifford at Armentières and Starzeene. Issued with several hundred Proto kits, he was now able to instruct men in large batches. Other former 'brigade men' assisted the training process. Clifford's work was appreciated so much that in June 1917 he was awarded the Meritorious Service Medal 'for services to the Army in the field' and promoted to Sergeant. There's no doubt that his input saved many tunnellers' lives. Clifford's career experiences took another professional turn when in January 1918 he was rushed home to aid the long aftermath of the terrible Minnie Pit disaster, a tragic event that resulted in the deaths of 155 miners.[56] By this time Clifford had been involved in the training of 3,000-4,000 officers and men in mine rescue on the Western Front.

Arthur B. Clifford continued to be involved in mine safety and rescue training for the rest of his working life, especially in the Forest of Dean. His *Rescue Man's Manual*, produced in 1922, was probably the first of its kind anywhere. Clifford's contribution to underground warfare provision during the Great War was enormous, as was his work at numerous mine disasters. But one wonders if he regarded his proudest personal achievement as saving miners' lives by teaching, demonstrating and promoting the use and importance of safety equipment rather than medal accolades. Clifford was a remarkably talented man from a talented family.

An array of Victoria Crosses

Fifteen 'miner-VCs' were awarded for actions during 1917, more than any previous year of the war. Four case studies are described below.

Lance Corporal Thomas 'Tom' Bryan (1882-1945), 25th Battalion (2nd Tyneside Irish), Northumberland Fusiliers [57]

Bryan's extraordinary act of bravery occurred on the very first day of the Arras battle (Vimy Ridge), near Roclincourt, on 9 April 1917. The Northumberland Fusiliers had come under heavy fire on the Thelus road. On an initiative with Captain Huntley to try and silence a very active and dangerous German machine-gun team, he opted to continue alone along a communication trench after his officer colleague had been killed, reaching the enemy position from the rear, where he disabled the gun having bayonetted two of its operators. The action took place despite incurring a gunshot wound to his right arm.

Evacuated to Blighty, Bryan was treated in Alnwick Military Convalescence

L/Corporal Thomas Bryan VC. Brian Elliott

Bryan's grave in Arksey Cemetery. Brian Elliott

Home. On 17 June 1917, a huge 'Toon' crowd of 40,000 witnessed the presentation of his VC by King George V at St James's football ground, Newcastle-upon-Tyne. Amazingly, little more than a year later, whilst still undergoing treatment at Norwich War Hospital, he rescued three-year-old Phyllis Richardson from drowning in the River Yare and managed to resuscitate her.

From Lye, near Stourbridge, in Worcestershire, Bryan was the eldest son of Thomas Bryan, a hewer who had settled with his wife Sarah in the Whitwood Mere area of the Yorkshire glass and coal town of Castleford when he was an infant. Thomas senior worked at Wheldale Colliery in 1911. Aged twenty-one, Tom junior married a local miner's daughter, Sarah Smart, on Boxing Day 1903, when he was then probably employed at Whitwood Colliery, Normanton, a pit owned by the Henry Briggs Company. A Castleford Northern Rugby Union player, he enlisted in the Northumberland Fusiliers Reserve Battalion initially, on 11 April 1915 and was in France by the year end, with the 25th Battalion. His wife Sarah was therefore left to care for their four children, which must have been a tremendous task for her and a heart-rendering experience whilst Tom remained on active service or in home hospitals (he was wounded or gassed three times).

Bryan returned to mining after the war, at Norton and Askern pits, near Doncaster and remained a face worker in 1939, according to his entry in the 1939 Register; but war ailments caught up with his arduous work underground. Consequently he set up in business as a greengrocer in the large colliery village of Bentley, a community which had been badly affected by the disaster of 1931 when forty-five miners lost their lives.[58] Thomas Bryan was buried in Arksey Cemetery, a CWGC gravestone marking the interment site. His VC medal is now part of the Ashcroft Gallery in the Imperial War Museum, London.

Sergeant Robert James Bye (1889-1962), 1st Battalion, Welsh Guards

Bye was a south Wales miner, born at Pontypridd but brought up in the village of Penrhiwceibr. He worked on the face at Deep Dyffryn Colliery, Mountain Ash, when he enlisted at the age of twenty-five with the newly formed Welsh Guards on 3 April 1915. Two of his brothers, David and Vivian, were already in army service and a third, Donald, was a boy soldier, enlisting at the age of fourteen. Married to Mabel Annie Lloyd in 1912, the Byes had one small child at the time, Robert Edward Bye, aged three, and a daughter, Jenny Bye was born during the war, in 1915.[59]

Robert was in France with the 1st Battalion

Former Welsh collier James Bye VC. Brian Elliott

from 18 August 1915. Fit, and of a muscular build (he was in the battalion boxing team), he progressed well through the ranks, his third promotion that of sergeant, on 4 April 1917. A survivor of the regiment's first major battle, at Loos, Robert's VC action took place on 31 July at the start of the Battle of Third Ypres (Passchendaele). The preliminary artillery barrage of the British left many German pillboxes unharmed. At Wood 15, north-east of Boezinge and north-west of Pilkem, it was Sergeant Bye who put one of them out of action, and then, whilst the Guards advanced, on his own initiative not only put paid to a succession of 'blockboxes', but led his party to achieve nigh unbelievable results. It was a truly extraordinary act of under-fire bravery and leadership. Bye was the first Welsh Guardsman to get a VC, and became a regimental and folk legend. Observers' comments that 'by his own hand' Bye had killed, wounded or captured over seventy enemy soldiers may not have been a complete exaggeration.[60]

Bye received a hero's welcome in the Aberdare Valley after his investiture at Buckingham Palace on 26 September 1917, though most of those who greeted him were living in considerable poverty. Discharged from duty on 1 February 1919, he subsequently re-enlisted with the Sherwood Foresters, serving until final discharge – on medical grounds – on 2 July 1921. After a spell of mine work in his home area of Penrhiwceibr, Robert migrated with his family to Warsop, in north Nottinghamshire, a coalfield area where opportunity and prospects were better than in old, now ephemeral pits of south Wales, finding employment at Warsop, Firbeck and Welbeck collieries. Remarkably, his military career re-started with the Sherwood Foresters again, as a Sergeant Major until medical discharge in 1941. His mining life, however, continued until 1955, retiring due to occupational (respiratory) disease.[61]

Acting Corporal Fred Greaves (1890-1973), 9th Battalion, Sherwood Foresters (Nottingham and Derbyshire Regiment)

Fred Greaves VC, a heroic survivor. Brian Elliott

Fred Greaves was a very modest man who was reluctant to recognise or discuss his outstanding military achievements, a fairly typical trait of a miner. From a Derbyshire family, Fred was born at Killamarsh, the eldest of twelve children. Aged twenty in 1911, Fred was worked as a 'filler', loading coal from the faces of the Staveley Coal and Iron Company's mines, initially at the family's local pit, Bonds Main Colliery. A little later, following his father and brother, he worked at Barlborough and Markham collieries, probably as a hewer. Like other miners, he experienced and witnessed a number of accidents but his were considerable, as he suffered a broken jaw and ribs in one and, at Barlborough, whilst still in his teens, was run over by a wayward coal wagon. His injuries were so bad that both of his legs were broken and his pelvis crushed. Greaves's rehab took a couple of years but determination was evident in his progression as a novice and then race-winning cyclist.

Initially rejected for military service due to his mining injuries, Fred's tenacity surfaced again when he was accepted with the Sherwood Foresters on 26 February 1915. Prior to his VC action he already had an eventful war, surviving relatively unscathed among many of his suffering mates at Gallipoli; and continued with the Foresters to Egypt, and then the Western Front, a few days after the Somme Offensive had started. Amid heavy fighting and growing casualties, Fred was wounded in the upper thigh of his left leg in the attack on Beaumont Hamel towards the year end. An embedded bullet was not removed until many years after his discharge but after hospitalisation in early 1917 he was able to rejoin his regiment on 21 May, as part of a Lewis Gun team. Fred's activities had caught the eye of his commanders as he was recommended, albeit unsuccessfully, for an MM. Among the subsequent family anecdotes about him, he was said to have displayed amazing courage when apprehended by Germans who, apparently, prodded him so hard with bayonets that his uniform was left 'in tatters'. Fred's escape was traumatic, as he was loudly instructed to keep still while one of his officers shot dead his German captors.

Fred's VC action took place on 4 October 1917 at Poelcapelle, east of Ypres, during the drawn out Passchendaele battle. After his platoon commander and sergeant were casualties, followed by another NCO, he rushed forward and bombed the occupants of a substantial concrete pillbox, killing or capturing the garrison and four enemy machine guns. This alone was an extraordinary achievement, especially when under fire. But later, at a critical period in the battle, and in absence of any offices capable of taking any action, he directed men on to the flanks and enfiladed (sweeping fire along the length of their troops) the German advance. Thus he was involved in two amazing, initiative-orientated actions on the same day.[62]

Fred was demobilised in January 1919, having been promoted to sergeant earlier and continued his work as a miner, as a deputy at Markham Colliery until retirement in 1955. He had an extra role at the mine, acting as a safety officer and St John's Ambulance representative whenever needed. And that certainly took place in a major way after the terrible disaster of 1938 when seventy-nine men lost their lives following an underground explosion.[63] Fred Greaves' heroism was well-marked locally in numerous formats before and after his death.

Private Albert Edward Shepherd (1897-1966), Rifleman, 12th (Service) Battalion, The King's Royal Rifle Corps (KRRC)[64]
A small, quiet, shy and unassuming young man transformed into an heroic giant in an extreme action on the Western Front in 1917. Later he returned to his civilian mining job 'without a murmur' of his achievement. The example

of Albert Shepherd is a classic. My paternal grandfather probably knew him as they both worked on the coalfaces at Monckton Main Colliery in the Barnsley area village of Royston. And my maternal granddad, also a Great War veteran, who found employment 'on the pit top' amid the coke and chemical works on the same colliery complex, became a close neighbour of Albert and his family. I remember a picture of 'Royston's war hero', wearing a dress suit, sporting a bow tie and displaying his medals in a large framed photograph placed above the hallway entrance to the Civic Centre, close to the secondary school where I taught in the 1970s.

Of a Shropshire mining heritage on his father's side of the family, the Shepherds tried several South Yorkshire locations before finally settling at Royston in about 1905. This large pit village already had a reputation for attracting many workers and their families from the West Midlands to its Monckton mining complex. Albert, aged fourteen in 1911, was employed as a pony driver and may have continued to work 'on the haulage' at the time of his enlistment into the 12th KRRC on 18 August 1915, aged eighteen. His mother, Laura, a local girl, had died suddenly, at the age of thirty-two, in 1909.

Maybe due to his youth and fitness (he was a talented boxer), Albert was deployed as a 'company runner', surviving the Somme and Passchendaele (Third Battle of Ypres). His VC action – almost unbelievable in its audacity and courage – took place near Villers-Pouich, northern France right at the start of the Cambrai campaign, on 20 November 1917, when the 12th with the 6th Oxford and Buckinghamshire Light Infantry and A Battalion The

Albert Shepherd, an ever-modest VC hero. *Barnsley Chronicle*

Plaque commemorating Albert Shepherd, Royston churchyard. Brian Elliott

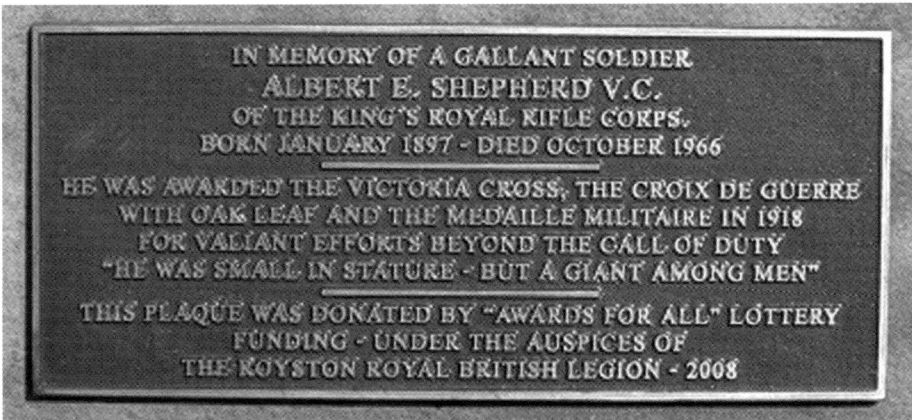

Stinking Trenches and Surreal Landscapes • 179

Tank Corps, had secured their first objective, reaching beyond the first three trenches of the Hindenburg Line. Attempting to advance to their second objective, Shepherd, still only twenty years old, volunteered to rush a German machine gun which threatened to annihilate his entire company. On his own initiative he then threw a Mills bomb, killing the two gunners and capturing the gun. When his officers and NCOs had become casualties he took command, ordered the men to take cover, and then went back some 70 yards – under fire – to procure the help of a tank, which providing cover for a successful advance. It was a truly remarkable, single-handed act of gallantry.

Shepherd's VC investiture took place at Buckingham Palace on 8 March 1918 but his homecoming, despite a 'no fuss' plea made via his father, proved to be an unprecedented occasion for the village. Albert's arrival at Royston railway station attracted a huge assembly of people. A more formal gathering followed at another palace, the Palace cinema, where worthies and dignitaries paid tribute to his achievements, with Albert being presented with a gold watch and chain. Celebrations, however, did not stop there as the very next day a relay of supporters carried him shoulder-high in a four-mile procession from Royston to the much larger town of Barnsley where a civic reception took place in the Chronicle Building overlooking Peel Square. Once again, afterwards he was 'publicly displayed' before the crowd, carried yet again on the shoulders of well-wishers. No wonder he wanted a quiet life subsequently.

Wounded twice and also gassed during his service, Albert Shepherd, now a Corporal, was discharged from the army in January 1919. He was soon to be the recipient of two prestigious French gallantry awards: the Croix de Guerre and Medaille Militaire. Albert's occupational life continued to be that of a miner at Monckton Main Colliery, his old pit, and he looked forward to a quiet life after he married Rosezillah Tillman in the parish church of St John the Baptist, Royston on 17 February of the same year. Unfortunately, a double tragedy followed, the death of their baby daughter and then of Rosezillah herself in 1925. Just over a year later he married Gladys Maud Lees and they went on to have five children, though two died in infancy. Albert was still working as an underground faceworker at Monckton in 1939, retiring a few years later after a spell of 'lighter duty' employment as a caretaker in the colliery offices.

The other VC holders having mining backgrounds, in chronological order, for 1917 are: **Michael Wilson Heaviside** (15/Durham Light Infantry); **Samuel Frickleton** (3/3 New Zealand Rifle Brigade); **James Llewellyn Davies** (13/Royal Welch Fusiliers); **William Boynton Butler** (17/Prince of Wales Own, West Yorkshires); **Wilfred Edwards** (7/Kings Own Yorkshire Light Infantry); **Thomas Woodcock** (2/Irish Guards); **Albert Egerton** (16/Sherwood Foresters); **John Harold Rhodes** (3/Grenadier Guards); **John Molyneux** (2/Royal Fusiliers); **John Collins** (25/Royal Welch Fusiliers); and **John McAulay** (1/Scots Guards).

***

Thus, many thousands of former miners, volunteers from 1914-15 – as well as the new targeted recruits – were kept very busy in theatres of war during the year 1917. It's no easy matter, but of the hundreds of case studies collected during the course of research, the following are selected to conclude this chapter, along with a note about diversity and a contemporary appraisal of their lasting contribution.

A teenage pilot killed at Arras

During what some military historians refer to as 'Bloody April', the Arras offensive, which began on 5 April 1917 was a land battle supported by the flying services of the British and French.[65] Carrying out vital reconnaissance and attacking missions, the information collected by the airmen was of strategic importance to ground forces. But in their actions the RFC suffered heavy losses, 275 aircraft and 421 casualties, almost half of them fatalities. One of these, Charles Sydney Hall, Second Lieutenant with the 60th Squadron RFC, whilst leading a fighting patrol of six aircraft on the evening of 7 April (two days after the start of the offensive), lost his life when his plane was shot down 'in front of Arras'. Only aged eighteen, under official combat age, his body was found two days later during a British advance. Hall was buried in the Tilloy British Cemetery, east of Arras. He was also a trainee mining engineer, the son of Joseph John Hall, a former colliery manager and mining engineer, the Hall family well-known residents in the Northumberland 'coal town' of Ashington. Teenage officers in the RFC – and indeed elsewhere – were given responsibility way beyond their life experiences, often at great personal cost.

Killed in the snow on the first day of the Arras battle

Private James Brown, a young miner from Ince in Makerfield, Wigan, was serving with the 10th Battalion, York and Lancaster Regiment when he lost his life on Easter Monday, 9 April 1917, aged twenty-five, during the First Battle of the Scarpe (battles of Arras). His local newspaper reported – surprising quickly, the next day – on the 'splendid British advance' that had taken around Arras, the capture of seven villages and Vimy Ridge. By two in the afternoon at least 5,815 prisoners were said to have been taken.[66] But no British casualties were mentioned in the local press. The regimental War Diary does indeed record the taking

James Brown's War Medal, with original ribbon.
Brian Elliott

of 12,000 prisoners over two days of action, hailing the action as 'the most successful battle yet fought on the British front'.[67] This was despite having to cope with 'heavy snow' at the start of the action. Casualties occurred at the start of the morning advance, after a German shell started a small fire in an ammunition dump under the steel bridge that served Arras railway station. The bridge was destroyed by 'a terrific explosion' that caused injuries and deaths to several men. Brown was one of the six immediate fatalities (several others dying of wounds afterwards), killed on the snowy morning of 9 April. In France from 27 August 1915, Brown was entitled to the Victory, British and 15 Star medals. He is remembered on the Arras Memorial and on the Ince War Memorial. He had married Margaret Ashcroft, a collier's daughter, also from Ince, on 28 February 1915 and their first and only child was born almost two months later.

A Gallipoli veteran loses his life at Arras

On 18 July 1908, 22 year-old Arthur Turner married Agnes Lee, two years younger, in their parish church, St Mary's, at Sutton-in-Ashfield, Nottinghamshire. Arthur was a miner, born just over the Derbyshire border, at Kilburn. Described as a hewer in the 1911 census, by the outbreak of the First World War the Turners had three small children, the youngest, baby Thomas, only a few months old.[68] For whatever reason, on 7 September 1914, along with several of his mates, probably from New Hucknall Colliery, where he had worked since leaving school, made the short journey to the recruiting office in Mansfield and enlisted in the Sherwood Foresters (Nottinghamshire and Derbyshire Regiment) but was transferred to the Royal Marine Light Infantry (1st Royal Marine Battalion) a few days later. In April 1915 Arthur was dispatched to Gallipoli where he was wounded and the enteric fever that he caught was so bad that he was invalided back to Britain to recuperate, a process that delayed his return to duty until March 1917. His convalescence must have been a difficult period for Agnes and Arthur, knowing that recovery would certainly mean a return to active service, at a time when she was expecting a fourth child. Before leaving for service, as a memento he gave her a beaded necklace which had a small locket containing a portrait of himself. It was the last time that they saw each other. On 26 April Arthur Turner was killed in action during the Battle of Arras. Two months later, anxious, grieving and still holding hope that he was still alive, Agnes sent him a postcard for his twenty-ninth birthday. Sydney, their fourth child, was born on 18 July 1917, on what would have been the couple's ninth

The poignant keepsake that Arthur Turner sent to his wife Agnes. Paul Atterbury

wedding anniversary. Private Arthur Turner, having no known grave, is listed among 35,000 Names of the Missing on the Arras Memorial. In Sutton-in-Ashfield, Agnes had a framed picture of Arthur as a lasting reminder of her late husband, the father of her four children.

Dai Jenkins, a Welsh 'ambulance man' in the RAMC

Many miners interested in helping the health and welfare of their workmates took an interest in First Aid, taking part in training with organisations such as St John Ambulance. Such men were ideal recruits for service in the RAMC. David (Dai) Jenkins, from Kenfig Hill, Bridgend, south Wales was the eldest of six children and like his father became a coalminer and St John's volunteer. Dai enlisted with the RAMC and as part of his initial training served in hospitals in the garrison town of Tidworth, at the edge of Salisbury Plain and also in the 'army town' of Aldershot, Hampshire. His first overseas service was in 1915, on a hospital ship in the Mediterranean, caring for war-wounded soldiers in the Dardanelles. His next posting was even more challenging, to Mesopotamia where so many soldiers suffered a variety of injuries and diseases in the Turkish war. On duty during the siege at Kut Al Amara, Dai was one of those captured after the British surrender on 29 April 1916. Fortunately, he was released, along with a party of 250 wounded soldiers, and transported to Basra, therefore not having to face the terrible fate of the rest of those that had surrendered. Promoted to sergeant, he continued to serve in Mesopotamia for the remainder of the war, the armistice with Turkey signed on 1 October 1918.

QM Sergeant John Stones: gassed

The wife and family of Quarter-Master Sergeant John Stones felt it right and proper that the circumstances his death were clearly marked under his name and regiment on his headstone. The inscription albeit partly faded can still be read today:

> WOUNDED IN ACTION (GAS POISONING) AT NIEWPORT, W.FLANDERS JULY 21 1917, AND DIED AT MILITARY HOSPITAL, CARDIFF, SEP. 2. 1917 AGED 31 YEARS.

The headstone wording concludes with a not-uncommon poignant phrase:

> HE DIED THAT WE MIGHT LIVE

A CWGC grave now marks the middle of his grave, in Thrybergh churchyard extension, Rotherham, its distinctive appearance standing out from neighbouring monuments.

The son of the Thrybergh village policeman, John lived nearby at Wickersley, working as a 'fitter' at Thurcroft Main colliery. A Rotherham Territorial from the age of seventeen, he married local girl Florence Green

The gravestones of John Stones, Thrybergh cemetery.
Brian Elliott

in 1908 and they had one daughter when he was mobilised after war was declared. On active service in France with the 1st/5th York and Lancasters in the spring of 1915, he was promoted within a few months, but his almost two years as a NCO came to an abrupt and debilitating end in a gas attack whilst he was overseeing supplies. Conveyed to Cardiff military hospital, his overall condition deteriorated when pneumonia developed. Thrybergh St Leonard's choir sang hymns in the church and at the graveside, and his coffin was carried to its last resting place by six of his former officers.[69]

Private Thomas Barrs, killed at Hooge

One of the most tragic aspects of researching the lives of the 'miner-solders' was the great number of volunteers who had survived intense, severe fighting only to die in later battles. Private Thomas Edward Barrs, aged twenty-four, serving with the 9th (Service) Battalion The King's Own Yorkshire Light Infantry (KOYLI), is a typical example, a young single man, killed in action on 4 October 1917 and buried in the Hooge Crater Cemetery, near Ypres (Ieper). Raised at Pontefract, west Yorkshire, the 9th was one of the Kitchener New Army battalions. In France from September 1915, it was involved in near-continuous fighting and a high number of casualties throughout the war. Barrs would have survived some of the most dangerous actions, from Loos and Arras and on the Somme when the 9th were almost wiped out. His fall occurred in the intense fighting around Hooge village and chateau in the autumn of 1917. The huge crater, close to the cemetery,

Private Thomas Barrs: his gravestone, portrait and War Medal. Brian Elliott

was a consequence of detonation by the 175TC in 1915. Employed as a miner at Wharncliffe Woodmoor Colliery, Carlton, he lived with his parents in Barnsley. Thomas had enlisted on 3 September 1914 and went on to serve with the British Expeditionary Force in France and Flanders. He was reported missing immediately after the fighting at Ypres, on 4 October, and locally several weeks later.[70]

***

Founded in 1912, the Royal Flying Corps (RFC) had five squadrons at the start of the First World War and about 2,000 personnel. From a fairly marginal support role, the RFC increased rapidly in importance, particularly with respect to artillery operations, photo reconnaissance and – from 1917 especially – in aerial combat missions. By 1 April 1918, when the RFC merged with Royal Naval Air Service (RNAS) to form the Royal Air Force, there were 150 squadrons, 114,000 personnel and 4,000 combat aircraft.[71] The officers and men drawn into the new service did so with little or no flying experience, usually transferred from infantry regiments.

Among the new and later transfers from the army were a number of 'specially selected' or 'volunteered' miners. Miners who worked as mechanical 'fitters', and in a variety of craft jobs involving metal and wood had skills easily adapted to working on aircraft maintenance duties; and there are some cases where this role transformed into flying in aircraft in combat and reconnaissance situations. Mining engineers, too, were well suited to aerial duties because of their technical skills and safety know how;

Stinking Trenches and Surreal Landscapes • **185**

and some were already well versed in the fast developing use of photography. The following examples, two from Yorkshire and one from Durham are but a small sample.

Ernie Antcliffe DFM: teenage 'flying ace'

'Ernie' (Ernest) Antcliffe's aerial exploits as an observer and rear gunner were so impressive that he was one of the first airmen to be awarded the Distinguished Flying Medal (DFM), and at only nineteen years of age one of the youngest.[72] His award was gazetted on the third anniversary of the medal's institution, given retrospectively for his actions during 1918.[73] Antcliffe was reported to have had 'fourteen months very active flying service in France' where he managed to avoid 'the great penalty of the heroism of his comrades' through numerous 'escapes'.[74]

The Antcliffes (or 'Anteclifes') were a Derbyshire mining family, drawn to the bigger pit villages of south Yorkshire, first Bentley and, from about 1910, on the concealed coalfield at Maltby where a large new colliery and community was in the course of early development on land owned by Earl Scarbrough of Sandbeck Hall. The Antcliffes would have been aware of the colliery owner's urgent need for skilled labour as they were also from Derbyshire, the well-known Sheepbridge Coal and Iron Company (SC&I) of Chesterfield. And a few years earlier, SC&I were also leaders in the opening of nearby Dinnington Colliery, where Ernest's father may have worked.

Still only seventeen years old, on 26 September 1916 Ernest Antcliffe left his home in Maltby, maybe with a pal or two, and made the short journey into Rotherham in order to enlist in the York and Lancasters. At the time his occupation at Maltby Main Colliery was that of a blacksmith's striker, so he would have been based on the pit top, with occasional duties underground. He had a deceptively skilled 'hot-metal' job, despite its predominantly physical requirements, making a variety of items for use in and about the mine. Placed on the army reserve, within a couple of days of his mobilization, on 14 February 1917, he was transferred from an infantry unit (270th Infantry Battalion) to the RFC, an indication that his skills were thought to have been far more useful in that area.

Shortly after his further transfer into the new RAF, on 17 April 1918, he was promoted to Sergeant, and re-mustered as Sergeant Mechanic two months later. The upgrade was due to Antcliffe's performances as an observer/gunner at the rear of a Bristol F.2 fighter plane in 88 Squadron. Actioning the gun involved – as the illustration shows – one of the most precarious and daring positions imaginable, a perfect image for Biggles stories created by Holt after the war. How successful a gun operator was Antcliffe? Well, his record was to say the least impressive: a total of seven aerial combat 'victories' against German Fokker D.VII aircraft. And he was only nineteen years old. Two of the Fokkers were 'shot down', totally destroyed by his rapid and accurate fire, two others were burst into flames midair and two more got hit so badly that they were driven downward, out of control. Three of the seven assaults were in a plane piloted by 88 Squadron's Australian-born flying ace Allan Hepburn, who was credited with sixteen victories in a glittering flying career.[75]

Pipe in his mouth, looking every bit a flying ace, Antcliffe's picture appeared alongside a short report about his award in his local newspaper, the *Rotherham Advertiser*, but thereafter details of his life are sketchy.[76] Family history research, however, indicates that a marriage, to Elsie Stokes, was registered in Rotherham

Operating a Lewis gun whilst standing on top of an aircraft must have been one of the most precarious tasks imaginable.
Paul Atterbury & Cambray family

Stinking Trenches and Surreal Landscapes • **187**

in 1922 and electoral registers show that he was resident in Maltby through most of the 1920s, so he may well have returned to find employment at Maltby Main. The family appears to have removed to nearby Swinton in the wake of the great 1926 lock-out of miners, and then on to Rawmarsh when in 1939, aged forty-one, his occupation is given as 'brick burner', residing with wife Elsie and son Kenneth, aged sixteen.[77] Ernest died in Worksop, Nottinghamshire on 19 June 1974, a somewhat forgotten flying hero of the Great War.

Lieutenant Maurice Nicholson: pioneer pilot, Killed in action at Arras

The son of a Sheffield steelworks owner-manager, Nicholson had different career aspirations to his father, having completed an engineering course at the Royal School of Mines, London, prior to the start of the war. A promising career in mining was interrupted and then ended by the war. Commissioned into the Norfolk Regiment whilst at Sheffield University, he was transferred to the Army Cyclists Corps, but opted to join the RFC, serving in France from 1915 'in a school of gunnery' when aged twenty-eight.[78] Pilot training at the time was often accomplished within a few weeks, and Nicholson was involved in aerial sorties within days of a return from leave. Reported as missing 'after a flight over German lines on 18 August';[79] his death was then confirmed in 'air combat over France' on 18 August 1917. Second Lieutenant Maurice Nicholson, 11th Squadron, Royal Flying Corps, is remembered on the Arras Flying Services Memorial, Faubourg d'Amiens Cemetery.

William Smith: RNAS/RAF seaplane mechanic

William 'Bill' Smith was an experienced miner, born at Shildon in County Durham in 1877. Shortly after the start of the war, in November 1914, he opted for service as in the Royal Navy. As a hewer he would have had a variety of skills easily adapted to mechanical uses and appears to have put these to good effect after he was accepted into the Royal Navy Air Service (RNAS) on 14 September 1916. In his fortieth year, Bill left a wife, Ann Isabel and several small children, for land-based mechanical training via HMS *President II*, based at Crystal Palace and Chatham and then from 2 October 1917 overseas at Dunkirk, France, transferring to HMS *Daedalus* on 1 February 1918. The RNAS was absorbed into the new RAF from 1 April two months later. Smith's service records show that he was involved in seaplane maintenance and repair duties whilst serving on *Daedalus*, a vital job for keeping the planes air worthy. He was indeed a skilled mechanic, promoted to Sub Lieutenant prior to demobilisation at the end of 1918 and transfer to the RAF Reserve. Unfortunately over the printed word 'Discharge' the clerical officer has written 'Deceased' and opposite the date 30 April 1920. Bill was one of numerous former miners whose abilities were recognised and used by the fast-developing air service. He was also one of many older miners who survived the war but died not long after discharge.

Appraisal

The great irony was that miners, especially in more radical areas, were 'red-baited', dubbed as 'dangerous to society' in the mainly conservative media – particularly later in the year 1917, in the wake of the revolutionary events in Russia. Local 'court reports' in coalfield newspapers reported magistrates punishing miners for minor offences such as 'neglect of work' and the oldest complaint of all: absenteeism.

South Wales, still smarting from the miners' strikes of 1915, continued to be regarded as *the* radical hotbed, a region 'riddled with distrust, suspicion and resentment'. The government's worry about 'Red South Wales' resulted in a local 'patriotic campaign' to allay 'radical elements'. It partly worked, as a subsequent ballot of union members resulted in a 98,948 v 28,903 vote *against* a strike if 'combing-out' was introduced. But the size of the opposition vote was significant.

Fearing more industrial unrest, on 29 October 1916 the government had resorted to DoRA in order to take control of the coalmines in Monmouthshire and south Wales, a precursor to the emergency nationalisation of the coal industry *as a whole* – amid growing 'disturbances in Russia – in March 1917. Three months earlier, speaking at the annual conference of the Labour Party, the president of the Miners' Federation of Great Britain, Robert Smillie, was all for the state control of the coal industry 'on the basis of production for use instead of production for profit'.[80] The government's action, however, was far removed from Smillie's vision, more a watered down, temporary control in which the coal owners continued to benefit. Nevertheless, a new post was created, that of Controller of Coal Mines (but responsible to the Board of Trade), headed by Guy Calthrop, assisted by one of the most knowledgeable persons in the industry, Sir Richard Redmayne, the Chief Inspector of Mines.

Writing not long after the end of the war, Sir Richard had this to say about the miners and the war in the final sentence of his landmark book, the insightful *The British Coal-Mining Industry During the War*:

> . . . whenever one is inclined to criticize them on the occurrence of strikes dislocating trade and industry, be it remembered that these are the same men who, when the safety, and indeed the very existence, of the country was in jeopardy, flocked voluntarily by hundreds of thousands to the Colours and did yeoman service in the Great War.[81]

There could not have been finer testimony, from someone who knew the miners and the industry so well.

Chapter Eight

BEHIND AND BEYOND THE WIRE

Newspapers published in coalfield areas make occasional references to former miners held in captivity during the First World War. It is not known how many were actually incarcerated but they appear to have formed a significant group in many camps (see outbox below 'How many prisoners'). There were peak periods when references to PoWs were frequent. News reports are particularly common during and after the retreat from Mons, and as a consequence of the German offensive in 1918. Less information about prisoners emerge from the 'trench-dominated' middle years of the war when mass captures and surrenders were less common.

A few first-hand accounts of what it was like as a prisoner of war appeared in print within a few years after the end of the war, but it was a long time before military historians gave much attention to the subject. In the late 1980s, using material deposited in the Imperial War Museum, Robert Jackson published one of the first popular accounts.[1] More recently, the work of authors such as

How many prisoners?

It is not known how many former miners were PoWs in the Great War and there are no official or published lists extant in The National Archives (TNA: The limits of First World War records). However, the TNA does state that an estimated 192,000 British and Commonwealth servicemen 'were taken captive'. An 'official figure' of 171,720 British officers and other ranks taken prisoner by Germany and her allies, more specifically 165,055 of them captured by Germany on the Western Front, represents about 6% of the entire British army.[4]

In view of the huge recruitment of miners into military service, especially during the 'rush to arms' period of 1914-15, it seems reasonable to say that somewhere in the region of 2,000 to as many as 5,000 ex-miners experienced some form of captivity during the war.

An assembly of British PoWs awaiting transport to camps in probably early 1916. Taylor Library

Richard Van Emden (2009), Heather Jones (2011), John Lewis-Stempel (2014), David Bilton (2016) and Oliver Wilkinson (2017), have helped to redress the omission, providing us with a better understanding and appreciation of life in and around German camps.[2]

From October 1917 to September 1918, the death rate of British PoWs was more than twice than that on the battlefields during the same period. A sobering sentence. Indeed, 'a Tommy had a better chance of survival in a Flanders trench than a German prisoner-of-war (PoW) camp'.[3] Treatment was certainly bad if the captive was a former miner made to work in German mines where employment conditions were far worse than at home, and the experience must have been particularly dangerous for 'non-miners' forced to labour for many hours underground in a variety of onerous jobs that they were not used to, such as hewing, filling and hauling.

Killed by a guard

At least 500 British prisoners are believed to have died directly as a consequence of beatings and shootings. Although this estimation represents a tiny proportion of the total of 165,055 British troops captured by the Germans on the Western Front, there also were 12,000 or so PoWs who died in custody. Disease, starvation and 'barbed wire disease' (neurasthenia:

'physical and mental disturbance') accounted for many inmates who were, in effect, 'murdered and buried in unmarked shallow graves'.[5]

Reports about abuse, including deaths of PoWs at the hands of guards, began to be reported in many of the coalfield area newspapers. Even allowing for propagandist bias, such information bears some credence. One south Yorkshire example stands out.

On the morning of 25 August 1918, a former Rotherham Main miner, Lance Corporal Oliver Card, of the 9th KOYLI's, walked to the latrines, about 120 yards from his German prison hut at Poilcourt camp, near Rheims, where he had been incarcerated since his capture three months earlier. A sentry approached Card whilst he was on his short journey and – for no apparent reason – commenced to beat him with a stick. The Corporal tried to protect himself with his arms, whereupon the guard stepped back and shot him with a pistol, later claiming that this action was due to 'retaliation'. However, the incident was witnessed by a pal of Card's, Private Steeples of the 7th Leicesters, who happened to have been exiting the latrines at the same time of the brutal assault.

Despite his terrible condition, Card managed to describe what had happened to him to a soldier from the East Yorkshire Regiment who had arrived at the scene just after the shot was fired. This man accompanied Card to a nearby hospital, where he died after two hours of agony.

An officer from the East Yorkshire Regiment, Company Sergeant Major R. Huit, wrote to Card's father with details of the fate his son. What an incredibly distressing letter it must have been for parents Charles and Maria to receive. Charles Card worked as a hewer, almost certainly at the same pit of his now dead son, whose occupation is listed as 'pony driving in mine' in the 1911 census. The Card family appear to have migrated from south Wales via Staffordshire, for Charles to find work at a Rotherham pit and build a new life in a south Yorkshire mining community. We only know about this blatant act of murder of a PoW and flagrant abuse of the Geneva and Hague conventions after Huit's letter was sent by Card's parents to the *Rotherham Advertiser*, and published on 18 January 1919.

Having received no response from complaints to the German camp commander at Poilcourt, it was anticipated that Corporal Card's case was to be cited when the Grand International Assize Court convened at Dover to hear evidence of barbarities committed against British prisoners of war in Germany.

Card's service record does not survive, but the *Rotherham Advertiser* report cited above suggests that he had joined the army at the start of the war, had had three spells of duty in France, during which he was wounded twice, prior to capture on 27 May 1918.

Typically, there was no post-mortem or investigation by the German authorities. Aged twenty-two, Card was hurriedly buried on the same day as his death, 25 August 1918, in a German military cemetery near the camp, his grave marked by a simple wooden cross. An escorted group of seven fellow internees are said to have partaken in a modest graveside service in his honour.

Forced labour

By 1916, most PoWs (up to 80% of the British troops captured by Germany on the Western Front) were deployed in 'forced labour' situations. For many – whether they were miners or not – it meant work in coal and salt mines. Mining was the most dreaded of all deployments. This was not surprising given that shifts in the dark could last up to fourteen hours, about double what was normally expected. Working, for example, in the great 'black hole' that was the huge Auguste Victoria Colliery, in the Ruhr, was so feared that self-inflicted injuries were known to have taken place in order to escape the 'uncontrolled tyranny' that reigned; and the men were subjected to beatings with shovels for not meeting output targets.[6]

One ingenious and courageous PoW avoided work in Auguste Victoria by hiding for hours in a dark corner of the mine, joining the end of his detail at the end of the shift. After his cunning ploy was discovered he countered his punishment and further work by sabotaging tools, adopted a 'go slow' approach and feigning 'plague' by daubing mustard, soap and salt on to his skin.[7]

The conditions that PoWs faced working in another 'black spot' coalmine, the Preussengrube, were 'appalling' according to eyewitnesses, with 'worn out' 'forced labourers' taken to hospital, some of them dying there, according to Corporal C.E. Green.[8] Accidents to prisoners working in coalmines were not uncommon, as at least sixty-two British PoWs are known to have been killed when working underground.[9]

As mentioned above, it was by no means unusual for non-miners to be dispatched to work underground. Twenty-nine-year-old Private Robert Stones of the KOYLI regiment, from Barnsley, was captured at Le Cateau on 20 August 1914. Stones was employed in his father's hay and seed business, but was forced to work as a miner. Exceptionally, he managed to escape from captivity, though not until February 1918, after more than three years of a hellish existence that he regarded as 'slave labour'.[10]

Some men openly refused to work because of the inhumane conditions – despite the expected consequences. Private David Lightfoot, another former Barnsley area miner, from Houghton Main colliery, was aged thirty-six, and an experienced hewer prior to his enlistment in the Northamptonshire Regiment. For refusing to work after capture and imprisonment, he was placed in solitary confinement on several occasions. Lightfoot endured eighteen excruciatingly long months in several German camps, most notably Lager Dulmen (at Haltern-Sythen), where he was forced to extract stone when working knee-deep in foul water. On discharge, he was one of the few who managed to get categorised as disabled. There's no evidence to suggest that he was fit enough to return to mining.[11]

From the winter of 1918-19, PoW 'welcome home' reports were frequent insertions in local newspapers, and sometimes featured a particular miner. This example concerns Corporal Harry Smith, a 'Scots Fusilier' from the pit village of Denaby Main, near Doncaster. Smith was a well-known local footballer whose war hardly got going as he spent four years in Mecklenburg and Schleswig camps after capture during the retreat from Mons. The news entry highlights his brutally cruel, inhumane treatment: 'He was "tied to the stake" for a fortnight for asking for hot water at the cookhouse; and he was confined in a disused bakehouse oven for another fortnight for refusing to salute a civilian farmer who employed him.'[12] Despite his 'famished state', Harry was forced to work for several months in a coalmine near Hamburg, having to endure fourteen-hour shifts.

Some of the early British PoWs were 'sent to the front', deployed on work activities close to 'action areas'; and used as 'hostages', veritable 'human shields', in the hope of putting off enemy air-raids.[13] The following 'forced-to-serve-on-the-front' and 'work-in-mines' case, not reported until 1919, was probably not exceptional.[14] Private George Bell of the 1st Lincolns, from Darfield, Barnsley, wounded in a leg and captured at Le Cateau on 26 August 1914, was interned at Doberitz camp, near Berlin. Bell was deployed by the Germans on the Russian front line where, unarmed and unprotected, he was placed under frequent hostile fire. There was little respite and an almost total absence of normal 'comfort' provided. Off-duty, he had to live in sub-zero conditions in an unheated tent, surviving on meagre or missing rations.

British PoWs pose before one of their huts at Doberitz Camp in 1918. Brian Elliott

Working twelve-hour days, many of his fellow prisoners died from starvation and exhaustion. Bell also had the ignominy of his parcels from home 'pillaged by the Germans', tins of condensed milk, accessed and then clumsily re-sealed.

In November 1917, George Bell was sent to Chemnitz, in Saxony, where he was subjected to yet more harsh treatment. His principled refusal to work down a coalmine resulted in a month's solitary confinement and bread and water rations. 'Frantic with joy' was how he described his and his pals' reaction to hearing about the Armistice. In celebration, windows of his hut were smashed, though compensation had to be made to the Germans prior to a 22-mile walk to Stettin, the first stage of a long journey home.

Released and ready for action again

Mexborough railway station in South Yorkshire was crowded with well-wishers, mostly mining families, on the evening of 3 July 1915 when the train conveying a former Denaby Main miner arrived from his military base at Aldershot. Private Taylor R. Collingwood, aged twenty, was on a two-week furlough after almost eight months of captivity in Germany.

A Denaby Main mines rescue brigade member, Collingwood was one of many 'ambulance men' fast-tracked because of their first aid and medical skills into the RAMC a few days after the start of the war. After only eleven days of attending to wounded soldiers in 'under fire' circumstances in Belgium, whilst serving in C Company of 22 Field Ambulance, his stretcher-bearing and aid duties came to a sudden halt. Collingwood's squad was captured by the Germans when attempting to rescue two wounded officers during fierce fighting at Ypres on 22 October 1914.

After his homecoming, an 'in his own words' report of his post-capture experiences appeared in the *South Yorkshire Times*.[15] Propaganda elements apart, Collingwood's experiences appear to have been typical of other British PoWs at the time. His initial capture was followed by an intimidating and brutal 'forced march' through German trenches: 'The [German] rank and file were insulting and abusive. They jeered at the prisoners, spat at them, and until checked by their officers, struck them with the butt end of their rifles.'

A horrendous three-day rail journey followed, to a new prison camp at the edge of Gottingen: 'We were packed into cattle trucks, forty-five men in a truck . . . terribly cramped and confined, without food, save for a few turnips that were flung into the truck'. On arrival at Gottingen, the prisoners were given coffee, 'poor stuff, without sugar or milk' and the next morning served with 'nasty black-looking bread'. Collingwood brought home a 'souvenir sample' of the single slice of bread issued to the men each day.

Taylor Collingwood was one of sixty prisoners crammed into a hut measuring 102 feet long by 16 feet wide. His first work assignment was 'to complete the construction of the camp'. Employed mixing mortar, carting bricks and 'generally labouring', he considered his treatment as 'very cruel':

> We were severely and harshly dealt with for the least thing. We were not allowed to smoke and those who broke this rule were punished by being strapped to a telegraph post, and kept there for four hours. I have seen this punishment inflicted in the dead of winter with three or four inches of snow on the ground, and then the poor victim has fainted long before the period of punishment was up.
>
> We worked hard and did our best, but sometimes in the early stages had difficulty in understanding the orders that were given to us. This used to make the Germans angry, and they would beat us with sticks or with anything they could get their hands on. I was myself often struck with the butt end of a rifle.

Collingwood described his initial incarceration as 'all sleep and work'. His only reading matter consisted of the *Continental Times*, or 'Continental Liar' as he dubbed it. Conditions gradually improved, though the food remained poor, many prisoners relying on parcels from home:

> The day's rations consisted of coffee at seven o'clock, and . . . bread . . . which the men demolished at one meal. For dinner there was soup, made either from preserved cabbage, which was sour and nasty, from a few potatoes and carrots, or from horse beans . . . The soup was served in basins . . . never more than half full. The food generally was utter rubbish, and there was no chance of lodging a complaint.

However, he praised the camp as 'the healthiest in Germany', having a doctor on site albeit only dealing with more serious wounds and illnesses.

Despite his ordeal, Collingwood expressed a strong desire to 'return to the front very shortly' and resume his medical duties with the RAMC. He had been a PoW in the early months of the war, from October 1914 to June 1915.

As in many coalfield area newspapers, the 'war columns' of the *Barnsley Chronicle*, included PoWs among casualty and 'missing' listings'. The following extract is typical, from a 8 June 1918 edition:

> Mr and Mrs C. Keaton, Pindar Oaks Cottages . . . have received a letter from their son, Pte C.E. Keeton, K.O.S.B. Regt. [King's Own Scottish Borderers],who has been missing since March 21st, saying he is wounded and a prisoner of war in Germany. Prior to the war he worked at Barnsley Main Colliery.

Letters from PoWs were also inserted on a regular basis, submitted by 'loved ones'. 'Mrs Jones', from the pit village of Carlton, near Barnsley received what she described as a 'welcome letter' from her brother, previously described as 'missing'. Private James Wright informed her that he was a prisoner of war in Germany and his meals were far from adequate. 'We get served with awful stuff to eat, including soups of all sorts of trash – fish bellies and black peas, some meat and some crocodile! They keep taking men in to the "garden" and put a cross on them and it's a wonder there are no more, considering the food we get!'[16]

Prison visits by miners' wives

Wives appear to have been allowed to visit captive husbands 'as a special privilege', provided that appropriate travel permits were obtained. Walter Parkes, a 27-year-old Company Sergeant serving with the 5th York and Lancasters, had worked as a hewer at two Barnsley area collieries, Monckton Main and Grimethorpe. A few days after the start of the Somme Offensive, he suffered from gunshot wounds that left his left eye 'blinded and his left shoulder bone 'splintered'. Reported as 'missing', on 9 July, Parkes' status changed to that of 'prisoner of war', one of several wounded men taken by the Germans. However, his actions prior to his incapacity, on 6 July, were so outstanding that his commanding officer recommended him for a DCM.

Parkes's award went through military officialdom, eventually gazetted on 22 September 1916, the citation as follows:

> For conspicuous gallantry during an enemy counter-attack. Arriving at a critical moment with a party carrying bombs, he grasped the situation, and, aided by a private of another regiment, he drove the enemy out of the sap. On a subsequent occasion he rendered valuable service with reinforcements.

Because of his internment, Parkes was unable to get his DCM until repatriation, which turned out to be quite a long process. After being given medical treatment by the Germans, along with other soldiers in a similar condition, he was transferred to Murren, a picturesque alpine village in Switzerland, 'for recuperation'.

Intent on going to see her husband, Parkes's wife Eva journeyed to London where she was met by 'lady guides' and a contingent of thirty other wounded soldiers' wives. The party then endured a hazardous channel crossing and a long railway journey to see their loved ones. Walter Parkes was a York and Lancaster (4th Reserve) Territorial, and was camped at Whitby when war was declared. As he was on the Western Front by the spring of 1915, according to his medal card, Eva would have seen relatively little of him since their marriage in the summer of 1913.

At Murren, the women were not only allowed to visit their men but permitted to lodge with them for two weeks. Mr and Mrs Parkes were even photographed there, the image and the story later appearing in a local newspaper.[17] What an extraordinary journey it must have been for Eva to see her husband.

After repatriation, Parkes was classed as part

Walter Parkes and his wife Eva, photographed at Murren in 1917.
Barnsley Chronicle

disabled on his discharge papers, therefore entitled to a small pension. He declared an interest in employment as a 'timekeeper' at his old colliery, a safer, far less strenuous job than he had before, though he would have been on much less pay than when he worked as a hewer. Like many thousands of former miners able to return to their old employments, it was on low pay, due to an inability to do physically hard work. Those that did continue, were soon subjected to a flurry of wage reductions and 'no-pay' periods during disputes.

Tunneller company PoWs

Information about PoWs from the tunneller companies of the Royal Engineers is sketchy, though several interesting examples from south Wales emerge thanks to the research of Ritchie Wood.[18]

Any illusions that twenty-year-old Hubert Aubrey Clifford had about service for king and country got a severe reality check in the trenches and tunnels manned by the 172TC at The Bluff, the strategically important heap of spoil located between St Elio and Hill 60, south-east of Ypres. Here, at the start of 1916, German bombs and bullets rained over the British front lines with such intensity that underground operations – in terribly wet and cramped conditions – were even more precarious than normal.

On 14 February, the exploding of a large enemy mine resulted in the death or capture of all of the forward trench workers, including Clifford. Goodness knows what Hubert felt at the time. Relief certainly, as he was a lucky survivor, but it must have been an emotion countered by utter bewilderment. Once again, his life was suspended on a thread of uncertainty. Maybe, too, he felt something of a coward, of 'letting the side down' – a far from unusual response to capture.

From Pentwynmawr, near Newbridge, in Monmouthshire, employed as a 'collier' by the Oakdale Colliery Company, Clifford had spent a year as a Territorial with the 1st Monmouths prior to embarkation to France with the BEF on 13 February 1915. The military importance of his mining background was soon valued as, on 4 August 1915, he 'attested in the field' into the 172TC of the Royal Engineers, mustering as 'tunneller's mate'.

A few days after being recorded as missing, on 19 February 1916, Clifford was reported to be at Giessen camp, on the outskirts of Frankfurt, part of a significant number of Monmouth men already in detention at this location. Earlier, a letter from corporals Ferris and Costin, published in the *South Wales Argus*, stated that as many as 120 'Monmouths' were 'Giessen inmates'.[19]

On the face of it, Giessen was a model prison camp. It had a church, church hall, good medical and recreational facilities and the inmates enjoyed, according to an extant 'bill of fare', a range of food and meals fit for any decent restaurant. Photographs in David Bilton's *Allied PoWs in German Hands 1914-1918* show the camp's apparent orderly and pleasant appearance.[20] However,

Bilton also provides a realistic account of what conditions must have been really like, quoting the experiences of a former inmate, Rifleman Hall of the 12th London Regiment. Hall was one of 580 Giessen prisoners dispatched in 1917 to work behind German lines. Less than half returned, most of them in a very poor state of health. Starvation and 'shot in cold blood' were not uncommon experiences, and many of the hospitalised returnees never recovered.

After repatriation, on 29 December 1918, Hubert Clifford was admitted to the King George Hospital, London suffering from influenza. His claim for disability status due to 'lung and chest trouble', requiring three spells in the Giessen camp hospital, was rejected by the army, perhaps on the grounds that his late condition occurred during the highly infectious second wave of the 'Spanish flu' pandemic. Accordingly, he was transferred to Class Z Army Reserve on 17 April 1919.

Clifford probably returned to mine work, albeit on a 'lighter' but less well-paid job as he appears on the 1939 Register, still resident in Pentwynmawr, as a 'general labourer', more specifically as an 'onsetter below ground'. Working at the pit bottom, he would have been well-known to all the men, as he was responsible for their safe entry or exit from the 'cages' towards the start and end of shifts. Well respected no doubt by his miner-mates and community, he managed to enjoy several years of retirement, passing away in 1974, aged seventy-nine.

From Penygraig, in the Rhondda Valley, William Rees Evans was employed at the Cambrian Combine collieries, Llywynypia, prior to enlistment into the Devonshire Regiment on 3 September 1914, aged nineteen. His stature and build was small, just 5 feet 3 inches in height and weighing under ten stones. Almost a year later, on 1 November 1915, he was transferred 'in the field' into the newly formed 253TC of the Royal Engineers.

The mining experience and skills of this Welsh miner became most obvious around the village of Hulloch, near Lens, northern France. Here, in April 1916, the 253TC and the Germans were involved in intensive shaft-sinking and tunnelling, in a noted coalmining area. The *London Gazette*, on 24 June 1916 records the award of a DCM to Evans, for 'conspicuous gallantry', in an action described as follows: 'When the enemy exploded a camouflet, Sappers Moore and Evans volunteered to enter the shaft, and, under very dangerous conditions, rescued, at great personal risk, no less than six men.'[21]

After remustering as a tunneller on 1 August 1916, Evans's war continued to be eventful as his service papers record him as 'gassed' in 1917 and as 'missing after capture' on 21 March 1918. On 14 December 1918, it was confirmed that he was in fact a PoW. After repatriation, like Clifford, he was unable to claim disability status, despite a fall in which he damaged one of his knees, and had a 'shortness of breath'. Transferred to Class Z Reserve on 25 March 1919, it is not known if he was well enough to return to mining.

It was far from easy for repatriated prisoners of war to get proper and adequate compensation for their resultant ill health. William Gardner, a 26-year-old collier from Pengam Garden Village, in the Rhymney Valley, probably employed by the Powell Duffryn Company, joined the 252TC on 12 February 1916. And what a shocking baptism it must have been. The 252 were involved in offensive and defensive mining during the fiercest of fighting on and under the Somme battlefield, most notably when creating and discharging the Hawthorne Ridge mine.

Reported missing on 20 November 1917, Gardner was declared to be prisoner of war six weeks later, on 1 January 1918, the news providing some hope for his young family. Diagnosed rather vaguely as suffering from 'Neurasthenia' (chronic fatigue and mental disability), he was assessed as 50% disabled by the army and, provisionally, awarded a pension of twenty shillings (£1) a week. A modest allowance of 14 shillings and 9 pence (about £0.75p) was allocated for his wife and three children.

Post-war treatment

After the Armistice there were numerous cases where repatriated former miners, clearly suffering from the physical and mental ill-effects of imprisonment, had their application for a disability pension turned down by the army. Even 'accepted' ones were often given reduced pensions. The authorities appear to have paid little heed to the condition of former PoWs.

An attempt to establish a national supportive organisation – a sort of database of PoWs – got sidelined and was soon forgotten. Perhaps that was a reason why PoWs were so neglected by military historians for so many years. They really were the 'silent witnesses' and 'lost men of the Great War'.[22]

It was often left to local, neighbourly goodwill for returned PoWs with little or no work prospects to get financial and subsistence help. Fund-raising for repatriated PoWs was a common activity in the coalfields, organised in the context of a single or group of collieries. In Yorkshire, and typical of other areas, a 'Comforts Fund' enabled thirty-five ex-Hemsworth Colliery miners to share the sum of £200 and the Denaby and Cadeby (collieries) Prisoners' Fund provided a treat and a small cash distribution to thirty-eight of its former workmen.[23] Concerts, cinema performances and a variety of events were organised in order to raise money for returning miners who had suffered so much in German camps.

Private Phillips's great escape

When 21-year-old Robert Phillips, a young miner at Elliot Colliery, New Tredegar, volunteered for service with the Welch Regiment at the outset of the First World War he had no idea what an extraordinary journey lay ahead. His

Private Robert Phillips. Lynda Osbourne story is of epic, film-making potential that lay untold and largely unknown until his granddaughter's research findings a century later. Robert's extant

Phillips photographed after his return to Wales on 25 December 1916, his parlous condition described as 'rancid, flea-covered and suffering from trench foot' by his son Glyndwr. Lynda Osbourne

diary, compiled after the war, formed the basis of amazing revelations, details of which Robert had kept quiet about in his unfortunately shortened later life.

In the trenches at the Western Front from early 1915, Robert soon experienced the hard, bloody and dirty aspects of war, most notably in May during 'Second Ypres' when the Germans used poisonous gas on a significant scale. A soda-soaked handkerchief covering his nose and mouth saved him from the worst of fates, but the sight of so many of his mates choking, dying and laying dead never left his mind.

Not long afterwards, amid ruinous Vermelles in northern France, he was wounded, captured and placed in several camps in western Germany until permanent detention at Homburg, which became his place of incarceration for fifteen long, painful months. Here, inmates were used as forced labour in a local coal mine. Initially, Phillips told his captors that he was a musician by trade but when the guards overheard him talking to a mate about mining his sham was exposed. Fourteen-hour shifts underground, in cold conditions wearing almost no clothes now framed his daily work. The supervisors were far from friendly as they were former German soldiers not long back from the front line who had been wounded earlier. In a letter home, Robert described how 'about a dozen of them would get onto one man, take him to a disused part of the pit and knock him about with their picks and kick him'.

An escape plan broached by some of his prisoner mates was abandoned as too dangerous but Robert was determined to go ahead even if it meant a

Behind and Beyond the Wire • **201**

solo attempt. He realised that the only weakness in camp security occurred when the guards changed shifts at night. In the minute or so when the exit was unattended he simply walked to freedom, one of the most audacious escape in the Great War. Carrying a few basic supplies and using survival skills, he travelled by night for several months, navigating by the stars through woodland and via countryside tracks, covering a distance of 200 miles until reaching the border with Holland. There he managed to avoid a patrolling German soldier by crawling to safety. After being helped by a Dutch family, Robert crossed the sea as a stowaway, eventually getting to Cardiff on Christmas Day 1916.

After news of Private Phillip's great escape reached New Tredegar he was given a rousing welcome, a thousand people attending an official reception, where he was presented with an engraved gold watch in commemoration of his 'daring and resource in escaping from Germany'.

However, Robert was far from well after his ordeal, suffering from 'the shakes' (PTSD), and damage to his lungs and nervous system due to his exposure to poisonous gas, so had treatment in a Yorkshire military hospital. After recovering sufficiently, he was – after a debrief – discharged from the army, but still placed on the Reserve.

Remarkably, Robert Phillips returned to mining but his state of mind must have received yet another shock when news reached him that his younger brother Eli had been killed on the Western Front in 1917. But there were happy times after his marriage to Mary Jane Howells in 1921, and the birth of a son, Glyndwr. However, in the wake of his terrible experiences on the battlefield, in prison camps and during his 'great escape', Robert's family life was tarnished due to depression and nightmares. Then tragedy struck in 1934 when he lost his life in a roof fall when working at Bedwas Colliery. Only forty years old, it was a cruel end for a remarkable survivor whose life has been thankfully uncovered and commemorated by his granddaughter Lynda Osbourne.[24]

The Leipzig War Criminal Trials and the Karl Heynen case

Initially, the Allies submitted a list of 853 names of alleged war criminals to the German government. But by the time the Leipzig Trials were held before the German Supreme Court, from 23 May 1921, only seven war crime cases were submitted by the British.[25] One of these concerned a notorious prison commander in charge of PoW labourers at the Fredrich der Grosse coalmine in the Ruhr, Sergeant Karl Heynen. A master cooper by trade, born in 1875, Heynen was well-known for brutality, dishing out a variety of cruel, uncalled for and totally illegal punishments to the miners under his responsibility.

Heynen was formally charged with using corporal punishment, including the use of his fists and his rifle butt, against as many as 200 British and forty

Russian PoWs. In addition, he was accused of treating a British inmate named Cross with extreme cruelty, throwing him in a shower bath and spraying him with hot and cold water for half an hour. In addition, he was alleged to have maltreated another British PoW called McDonald, hitting him with his rifle butt, knocking him down and kicking him, after the latter had attempted an escape. On 14 October 1915, Heynen was also charged with threatening to 'summary execute' the miners under his command who were 'on strike' – unless they returned to work.

Previously court-martialled by the Germans, and convicted of the main offences cited above, a very modest sentence of fourteen days detention – suspended until the end of the war – had been given to Heynen. This was 'set aside' by the Leipzig court at the insistence of the British government, ensuring that Heynen was placed before the war crimes judges. Surprisingly, the judges acquitted him of his actions and his threats during the strike incident, ruling that PoWs were entitled to complain – but not to refuse to follow orders. However, the justices did find Heynen guilty of fifteen other instances of brutality, mostly relating to the miners. He was given a custodial sentence of ten months in a civilian prison.[26]

Incoming verbal accounts of malnourishment and maltreatment of repatriated PoWs must have shocked the troops who greeted them at arrival places such as Dover, Hull and Leith prior to their local homecomings. Words such as 'emaciated' and 'like skeletons' were commonly used to describe the men's appearance after disembarkation.[27] Repatriation actually dragged on for many agonising months, the last PoWs not freed until March 1919.

The Reception Centres, for example at Ripon and Canterbury, included recorded interviews with prisoners regarding how they were treated as well as their conditions of incarceration; and, subsequently, a governmental report on 'forced British labour in the coal and salt mines' found widespread evidence of abuse. This contributed to the government's apparent determination to seek justice at the Leipzig War Trials. It was all too little and to no avail. Unfortunately for so many, by 1923, 'war crimes' were of little importance to a government preoccupied with post-war international politics and domestic reconstruction.

The former PoWs were doomed to be the 'forgotten heroes' of the First World War. Perhaps the most forgotten of the forgotten were those men, including former miners, who had to endure the wrath of their Turkish guards, only about half of them surviving incarceration. In the spring of 1916, the forced 'death march' towards imprisonment in Aleppo – after the Siege of Kut – remains one of the most blatantly cruel acts of the First World War.[28]

The Heynen case remains a unique instance of a German prison officer found guilty of cruelty to the miners under his command.

Chapter Nine

1918: DARK PITS OF WAR NO MORE

*'I thought of some who worked dark pits
Of war, and died...'*
Wilfred Owen[1]

As the war entered its fourth full year former miners were dying on and below foreign fields, whilst their collier mates at home were losing their lives in the coalfields of Britain. The human cost of getting coal for domestic and war fronts remained staggeringly high. By the year's end, 1,487 lives had been lost in fatal accidents in coalmines, or about one death every six hours.[2] New and young entrants to mining were particularly vulnerable to mishap, lacking the skills and physical strength of experienced men.

On Sunday 12 January 1918, the Minnie Pit of Podmore Hall Colliery in north Staffordshire 'fired', resulting in the deaths of 155 men and boys.[3] Almost one in three of the fatalities were teenagers, forty-four of them under the age of sixteen. Local rescue teams made heroic attempts to save lives, the captain of the Birchwood Colliery Number 1 brigade losing his own life in the process. Such was the

Rescue team at Minnie (Podmore Hall) Pit, Staffordshire, 12 January 1918. Fred Leigh

Wilfred Owen (1893-1918) and the Minnie disaster

The Minnie mine disaster occurred when Owen was stationed with his regiment, the 5th Manchesters, in Scarborough, only five weeks after receiving treatment for shell shock at Craiglockhart War Hospital, Edinburgh.

Clarence Gardens Hotel was used as the officers' seaside billet, and Owen – placed in charge – occupied a turret room with a commanding sea view. It was from his desk there that he was moved to draft 'in half an hour' the eight-verse poem *Miners*. Owen's empathy for the victims of the disaster may have been a product of knowing what former miners were like in war situations, many of the miner-soldiers under his leadership having worked down Lancashire pits. However, the poem also encapsulates his own recent and very frightening experiences on the Western Front, including 'being blown in the air' and falling into 'dark cavities' during German artillery attacks. Later he admitted in a letter home that the composition got 'mixed up with the War at the end'.[4]

Wilfred Owen in uniform.

Returning to France in August 1918, Owen was allocated a prestigious officers' gallantry award – the Military Cross – for an action at Joncourt on 1 October. Sadly, he was unable to see the publication of the citation after the war as he was killed in action whilst leading soldiers under his command to cross the Sambre-Oise Canal, near Ors, on 4 November. This was only a few days before the signing of the Armistice, the heartbreaking news reaching his parents on Armistice Day itself.

Miners appeared in *The Nation* on 26 January 1918, his first poem in print nationally, and one of only five others published in his own lifetime.[5] Verse four (of eight) is very powerful in relation to the men and boy miners:

> But the coals were murmuring of their mine,
> And moans down there
> Of boys that slept wry sleep, and men
> Writhing for air.

Wilfred Owen's reputation as a Great War poet widened afterwards, thanks to his compositions being compiled by his friend from Craiglockhart, Siegfried Sassoon. A decade later – with more background – another war veteran and literary figure, Edmund Blunden published his work.

scale of the explosion that it took nineteen months for Podmore to get back to normal, the restoration under the experienced guidance of mine rescue instructor Arthur Clifford, transferred from the Western Front by the Home Office to assist with the operation.

The impact of the Minnie disaster on mining families in Halmer End and neighbouring pit villages was enormous, the area already reeling from losses of loved ones in the war. Although the colliery company helped with some financial aid to the widows – as did the miners' unions and a public relief fund – the main source of solace for the bereaved were from neighbours and friends. The parlous situation of the bereaved, however, was made even worse because of the increasing effects of rationing, indeed for some larger households the loss of one or more main wage earners plunged them into destitution.

Miners in demand – again

The demand for coalminers continued unabated. Most colliery managers strove to keep production going to previous levels by employing workers from all walks of life, including school leavers as young as thirteen. The Government's thirst for 'military miners' was demonstrated when a further 50,000 recruits were 'agreed' (via volunteering or ballot) to be released, in addition to a quota of the same number already in the course of recruitment. So much for 'mining' being a 'reserved' occupation post 1916.

However, colliery managers were beginning to feel the benefit of some of their former workers returning from the army in increasing numbers. Thus by the end of 1918, well over a million workers were employed in the mines (1,029,688), only 1.2% short of the figure for 1917. Surface workers – vitally important for the operation of all collieries – numbered 222,602, a much harder figure to maintain than underground labour as from 1916 most of them were liable to compulsory enlistment. These included pivotal employees such as winding engine drivers and 'corporals' (master or foreman hauliers) and the pit-top hauliers themselves, all of whom were very hard to replace.[6]

About one in twenty (11,761) of all surface workers were female in 1918, a significant number bearing in mind their high concentrations in Lancashire, Scotland and Cumbria – but total exclusion elsewhere.[7] 'Pit-brow lasses' were often used on a variety of tasks such as hauling railed tubs of coal to the stacks and skips, sorting coal on the screens; and working as weigh cabin operatives, storekeepers and timber stackers. Almost a thousand of female mineworkers were aged thirteen to sixteen, such was the desperate need for labour.

Despite the great size of the workforce, production targets often dipped well below expectation. In south Wales, vital for its quality steam coal for the Navy, mines inspector Dyer Lewis reported that annual production fell by almost two million tons in 1918, and was seven million tons short of the prewar figure.

Keeping together a buoyant but inexperienced workforce did not satisfy the huge domestic and war demand for coal, even where, for example in Yorkshire, some pits produced coal on Saturdays, normally 'an off day'. Another factor that hindered coal extraction was that work on the sinking and development of new collieries had been suspended or slowed during hostilities. Maintaining coal production also required a far bigger investment by the private owners in the mechanisation of existing collieries. Thomas Mottram, a veteran Yorkshire mines inspector, concluded that the war showed the great importance of 'machine over hand labour and horse power'.[8] East of Scotland collieries production suffered as the majority of 'those taken by the military' were 'drawers' (hauliers), affecting the underground transport of coal; and sections of pits had to be 'stopped' (sealed off) and double-shift practices ceased.[9]

The Military Service Act of 1916 was tweaked hurriedly and the ever hard-pressed Colliery Recruitment Courts had more difficult decisions to make as to who was or was not eligible to be called up. The 'de-certification' of large cohorts of men 'on occupational grounds' who had entered mining since August 1914 made decisions difficult and was inconsistent between regions.[10] William Straker, secretary of the Northumberland Miners' Association, complained and drew attention to the administrative complexities, colliery managers having to complete forms for employees who were unmarried or widowers without any children (providing they were eighteen years and eight months old in but under the age of thirty-two on 1 May 1918). The documents had to be counter-signed by a checkweighman or a person of a similar status. Each mine and region was required to meet a fixed quota. More pressure, and greater bureaucracy ensued. The desks and voices at pit offices and recruiting courts groaned and grumbled yet again.[11]

The desperate need for more miners for the war fronts was at odds with the views of some union leaders who, since at least 1917, had been advocating a peace settlement. Delegates of the South Wales Miners' Federation went so far as to pass a resolution that the government should 'seize every opportunity for securing peace' – in line with the Labour Party's 'war aims'.[12]

Also in March, under the aegis of the MFGB, a ballot was approved for its regional associations – seventeen in all – for and against the new 'comb-out'. All, apart from the relatively small Bristol and Forest of Dean areas, organised a vote, so it was a fairly comprehensive gauge of the feelings of the miners nationwide. Only two associations voted in favour of the comb-out: Northumberland and Durham, though the Council of the latter had voted in support a few weeks earlier.[13] However, the majority, 219,311 for and 248,756 against (6.1%) was hardly a convincing mandate for opposition. Of the major coalfield associations, only Lancashire had substantial support against the comb-out. Consequently, the executive of the MFGB, after a meeting with Prime Minister Lloyd George, decided to 'take no action upon the ballot', saying that there would be no 'downing of tools'.[14]

The mood was certainly not right for strike action. Dyer Lewis, the south Wales mines inspector, referred to a 'great wave of war enthusiasm [for further enlistment] through the coalfield'. In the north of Wales, young miners were already 'offering themselves in large numbers for enlistment' without waiting for the ballot result. Recruiting offices were said to be 'overwhelmed'.[15] In Nottinghamshire, recruiting places were said to be 'besieged by young miners'.[16]

But matters came to ahead in May when the national conference of the MFGB, chaired by the prominent 'anti-conscription man' Robert Smilie, assembled in London. Its principal purpose was for the 139 delegates to hear and respond to the views of the Coal Controller, Guy Calthrop. Not unexpectedly, Calthrop appealed for 'regular work' (and less 'absenteeism') and 'increased effort' in order to boost output, though he did concede that the industry was under great difficulty due to the prospect of losing 100,00 men to the military. Delegates responded in no uncertain terms that taking out men under the age of thirty-two meant the loss of the strongest and most productive of the hewers who got the coal. The conference was said to have 'collapsed in disgust' accordingly.[17]

Rationing

Sugar was rationed in January and by April basics such as meat, butter and margarine followed. The process, which was to guarantee supplies rather than reduce consumption, was well organised and regarded as successful as it maintained a healthy diet for all, poorer people actually benefiting from a 'better share' of quality food, it was claimed. Queues outside shops and empty shelves did not materialise. Mining families were used to producing home food via allotments and smallholdings. However, reports showed that some families struggled. In south Wales, Merthyr miners convened a union meeting, Noel Ablett demanding that the government should distribute food *equally* among the districts.[18] The South Wales Miners' Federation received a deputation from the Pontypridd and Rhondda areas, complaining that the shortage of food was so serious that miners had insufficient 'snap' to take with them to work. Neighbouring areas also experienced similar problems, so Lord Rhondda was approached for intervention and action.[19]

The rationing and home economy schemes also included the early closure of shops in order to conserve energy (principally coal) and there were reduced hours of opening for pubs and clubs. Coal was also 'rationed', households limited in the use of fires *per room*, and hot baths were discouraged, the latter hardly helpful for miners returning home 'in their pit muck'. But, again, mining communities were ahead of others, able to obtain coal via 'coal picking' on the colliery waste heaps, a well-known activity during hard times.

Miners and the War

Miners continued to be active in all theatres of the war. At the start of the German Spring Offensive, on 21 March, through to the 11 November Armistice, they participated in battles in Belgium, northern France, Italy and the Middle East.

On the battlefields of the Western Front the role of British miners as tunnellers continued to evolve. At the start of the German offensive they were as active as ever, organising, supervising and creating new defences, and became more directly involved with fighting. The tunnellers of the Corps of Royal Engineers became part of a more integrated front that helped win the war, combining their special skills with those of their Canadian, Australian and New Zealand counterparts.

Thomas Parry: a veteran Welsh tunneller and his son, Daniel

The Parrys were a Welsh family settled for a while in Risca, Monmouthshire, where the Black Vein Colliery was the major place of employment for men and boys. Later, they moved to the small Rhondda pit village of Ynyshir. Here, in 1901, 27 year-old Thomas Parry, his wife Martha Ellen and their two small children Daniel (aged four) and Blodwen (six) shared a property rented by Henry and Elizabeth Philipps, Martha's parents. Henry and Thomas were miners at 'Lady Lewis', a flourishing colliery which had a workforce of about a thousand at the start of the First World War.[20] In 1911, following the decease of Henry Philipps, Thomas headed a small household consisting of his wife Martha, their two teenage children, mother-in-law Elizabeth and a relative of the latter, residing as a boarder. Thomas worked as an 'examiner' (a fireman or safety official) at the colliery and son Daniel, although only fourteen, is listed as a 'hewer' on the census form.

Just over a year after the start of the war, on 8 August 1915, a few months after his son Daniel had joined the 1st Dorsetshire Regiment, Thomas journeyed the twenty or so miles to Cardiff's recruiting office and was enlisted into the Royal Engineers as a Sapper, in the new 171 Tunnelling Company. Like many of

Lady Lewis Colliery, Ynyshir. Author's collection

> his miner mates, he was of a fairly small stature, standing 5 feet 6 inches and weighing 10½ stone. He gave his age as thirty-nine, though he was actually two years older, perhaps a ploy to make sure that he was accepted. Father and son were shipped separately to France after only a short period of training at their respective bases. Thomas's pay was enhanced a little further above that of a basic infantryman, when he re-mustered as a tunneller, from his original status of that of a 'tunneller's mate'.
>
> Father and son spent most of their service on the Western Front in highly dangerous battlefield situations. One wonders if they were ever able to communicate despite having contrasting roles. Although Parry senior missed the major offensive of the 171TC in April 1915, at Hill 60, he was subsequently active in offensive and defensive mining, most notably at Houplines, La Touquet, Chappel a Armentières and Ploegsteert. Parry junior was 'in the thick of it' too. The Dorsets were the first British unit to face a German gas attack, on 5 May 1915, also at Hill 60, an action that Daniel may have just missed but he was certainly busy on front lines afterwards – on the Somme, at Messines and during 'Passchendaele'. A few weeks after the start of the German offensive, on 11 April 1918, he was fatally wounded. Sadly, only the day before, the London Gazette announced that twenty-two-year-old Private Daniel Parry had been awarded the Military Medal. Private Daniel Parry MM was buried at Wimereux Communal Cemetery, France. News of his death must have been a great blow when it reached his father and family. By then Thomas, was a veteran serving in the 173TC, and had been hospitalised in 1916. Nevertheless he had returned to service, surviving numerous actions in the Ypres Salient at the time of his son's death and survived the war. Understandably keen to get home, Thomas did not claim to have any disabilities when completing his demobilisation statement, though having had kidney problems and enduring so many years of 'tunnelling', it was unlikely that he was able to return to his old mining job. Inscribed in upper case letters at the base of his son's grave are the words: A LOVING SON A BROTHER KIND A BEAUTIFUL MEMORY LEFT BEHIND, perhaps knowing or seeing this, it was perhaps some solace for the Parry family.[21]

Amid a flattened, battered landscape after Third Ypres (post November 1917) – in order to maintain their position – there was little the British could do other than 'dig deep' and 'go subterranean' so as to maintain a presence in the Salient. Accordingly, the Allied Command deployed many thousands of 'miner-tunnellers' – supported by detached infantrymen – into the area. By the spring of 1918 about 200 'independent' (but interconnected) underground shelters were created, an extraordinary achievement. The big ones, at Wieltje and Hill 63 had capacities to hold hundreds of troops. It was said that more people lived underground around Ypres than the population of the entire town![22] By autumn, tunnellers had built more than 370 dugouts in the region, sheltering between fifty and 2,000 men up to 10 metres or more below ground. Damp, foul-smelling

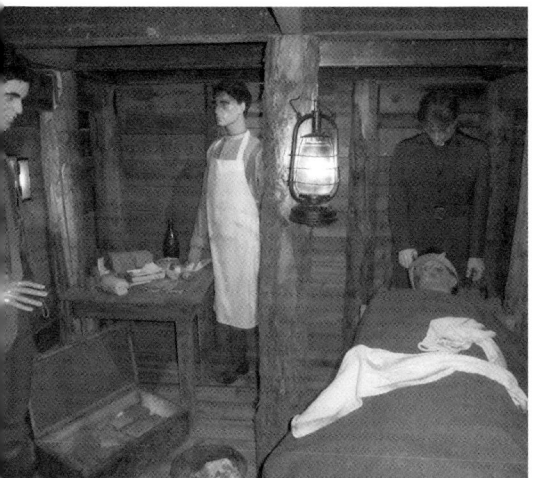 

Dugouts recreated at the Passchendaele Memorial Museum. Brian Elliott

and loved by vermin, they were however relatively safe places, providing reassurance for many.[23]

The re-discovery in 2007-8 of the huge underground shelter known as the 'Vampyr dugout', near Zonnebeke village confirmed the skills of the tunnellers, specifically the 171TC.[24] The huge shelter and its networks, which had the capability to house a brigade headquarters and all its requirements, was created almost 50 feet (fourteen metres) in little over four months. Operational from April, it soon fell to the Germans in the Lys battle; but was recaptured a few months later, in September 1918, the 2nd Worcestershires its final residents. It is displayed in model and reconstructed form in the Memorial Museum Passchendaele (Zonnebeke).[25]

The amount and range work that the miner-tunneller and engineers carried out in the last few months of open warfare was a remarkable testimony of their role towards the closure of the war. A statistical summary of their achievements was itemised a few years later as follows:

- construction of 149 heavy bridges
- construction of 38 light bridges
- removal of 6,714 enemy land mines
- removal of 315 delay-action mines
- removal of 24,725 demolition charges[26]

Haig's grateful tribute

The eventual withdrawal of many tunnellers from the army 'for urgent work at home' brought this glowing praise from Sir Douglas Haig:

Before they leave the country I wish to convey to the Controller of Mines ... my very sincere appreciation of the fine work that has been done

British tunnellers at work underground. Hill 60 info-image: Brian Elliott

by the tunnelling companies through the last four years. At their own special work, mine warfare, they have demonstrated their complete superiority over the Germans, and whether in the patient defensive mining, in the magnificent success at Messines, or in the preparation of for the offensives of the Somme, Arras and Ypres they have shown the highest qualities both as military engineers and as fighting troops. Their work in the very dangerous task of moving enemy traps and delay-action charges on subways, dugouts, bridging roads and the variety of other services on which they have been engaged has been on a level with their work in the mines. They have earned the thanks of the whole Army for their contribution to the defeat of the enemy. Their fighting spirit and technical efficiency have enhanced the reputation of the whole Corps of the Royal Engineers and of the engineers of Overseas Forces.[27]

Miners as VC war heroes

A huge number of heroic deeds were enacted by former miners during 1918, seventeen of them getting the highest and most prestigious award of all, the Victoria Cross. The last of these was won by **Sergeant William McNally**, a former pit pony boy from Murton, near Seaham, County Durham, serving with the 8th (Service) Battalion, Yorkshire Regiment.[28] His VC action took place in Italy, in late October, at Tezze (River Piave) and Vazzola, almost two weeks before the signing of the Armistice. On the 27th, at the Piave river, when his company was seriously threatened by machine-gun fire, he rushed an enemy post single-handedly, killing the crew and capturing the gun. In a similar action two days' later – when defending a recently captured ditch – he captured yet another German machine-gun post, at Vazzola, dispersing the enemy and inflicting numerous casualties in the process.

William McNally VC.
Public Domain

Remarkably, although still only twenty-three, McNally already had a notable record for bravery, awarded the MM for an action on the Somme on 10 July 1916 for dragging a badly injured officer to safety. Then in early November, before departing for Italy he got a Bar to his MM when on three occasions he rescued injured and buried men following artillery fire at Passchendaele. His actions were all the more remarkable given that he had suffered two or three leg wounds serious enough to merit medical care and home leave on two or three occasions.

After discharge from the Army in 1919, despite old injuries that must have still been extant, McNally continued his employment at Murton, retiring at the age of sixty-five, having latterly worked on the pit top as a timber-yard foreman. He passed away in his early eighties in 1976. A memorial tablet (of stone) was placed in Murton Park two years later and in 2018 – as part of the centenary of the war – a commemorative paving stone was laid by the cenotaph, one of seven for the County of Durham's 'VC heroes'.[29] McNally's Victoria Cross was the *fiftieth* awarded to miners during the Great War, equivalent to almost one in four of all British and Commonwealth recipients, a measure of how important miners were during the Great War.

At only nineteen years old, **Lance Sergeant Edward Benn ('Ned') Smith**, residing with his parents Charles Henry Smith (a fisherman) and Martha Smith, at Netherhall, Maryport, in Cumberland (Cumbria) was the youngest recipient of the award in the entire war.[30] Ned was employed as a pony driver by the Oughtibridge Coal Company's colliery, a relatively small mine employing about 465 men and boys in 1915, when he enlisted with 1/5th Lancashire Fusiliers.

Smith's 'double action' took place on 21/23 August, at Beauregard Dovecote, east of Serre, during an early phase of the Allies' Hundred Days' Offensive. In charge of a platoon, he captured a machine-gun post at 'The Lozenge' (Hill 140) and, ignoring an onslaught of grenades, proceeded to shoot at least six German soldiers. Responding to a counter-attack the next day, he led his soldiers forward and restored part of the line. 'His personal bravery, skill and initiative were outstanding, and his conduct throughout an inspiring example to all', according to a supplementary post in the *London Gazette*.[31]

Like McNally, Smith had won a previous gallantry award, his heroic deed taking place only ten days earlier (during the Battle of Amiens) when in charge of a daylight patrol near Hebuterne in the Somme area. Then a Corporal, he spotting about forty Germans intent on starting a night outpost position, and although well outnumbered, led his patrol forwards, 'inflicting heavy casualties and scattering them'.[32] For this action he received the DCM and was promotion to Lance Sergeant. Sadly, Ned Smith 'VC hero' died of a gunshot wound to the head (possibly from 'friendly fire') at Bucquoy, France on 12 January 1940, aged forty-one, whilst in service with the Lancashire Fusiliers.[33]

Sergeant William 'Bill' Gregg and **Private William 'Bill' Beesley**, serving with the 13th Battalion, The Rifle Brigade (Prince of Wales's Own), got their VCs during the Spring Offensive for differing deeds on the same day, 8 May 1918, at Bucquoy.[34] William Beesley's extraordinary deed of gallantry took place after his sergeant and section commanders were killed by enemy fire. Taking control, he rushed the offensive post, shot four of the enemy gunners, and managed to take six prisoners, returning them to his lines. Then with the help of a fellow private, Bill used his Lewis gun and inflicted further enemy casualties. When his comrade was wounded, he carried on holding his position until nightfall, returning the injured man and the gun to the original line, the latter 'kept in action' until 'things had quietened down'.[35] It was an astonishing series of acts that Beesley

Beesley and Gregg, miner-VCs. Daily Mail

performed over the course of almost a whole day, and hard to comprehend for someone who was only 22 years old.

Born at Linton (Church Gresley) in south Derbyshire, at the age of fifteen Beesley, was resident in the household of his stepparents James and Emma Ealing, in the Warwickshire village of Ansley, near Nuneaton. James was a 'holer', undercutting seams of coal with a pick (and other hand tools) on coalfaces, while William was employed as a 'point lad' on the 'pit bank', probably loading coal close to the main colliery buildings. William had transitioned to underground work at the large Haunchwood (and Tunnel) colliery complex at the start of the war, probably as a faceworker or haulage lad.

Beesley, probably alongside numerous of his workmates at Haunchwood, had been keen to enlist in the army, going along to the recruiting office – Nuneaton Police Station – a few days after war was declared, but as he was still only eighteen years old his application was rejected. Undeterred, he returned and was accepted into the King's Royal Rifle Corps, after giving his age as twenty. By June 1915, Private Beesley was in action with the 9th Battalion in the Ypres Sector. Two woundings followed. The first, in July, was to his shoulder, when hit by a piece of shrapnel; and then in November he suffered gunshot wounds to his legs when taking supplies to the Julien road. Back on the Western Front after hospitalisation, he formed part of a machine-gun team during the Somme battle.

Demobilised in 1919, Beesley returned, albeit briefly, to mining and married a year later, settling in Coventry. He furthered his military career when he joined the Royal Artillery at the start of the Second World War, finally leaving the army in 1942 on account of his age. Subsequently, he worked as a commissionaire for an engineering company until retirement in 1960. He died in hospital, aged seventy, after being taken ill on holiday in south Wales.

The bravest miner-soldier during the Great War? Well of course it's pointless to name one as there were so many heroic deeds. One name, however, certainly does stand out because of how he was regarded by contemporaries during offensive and defensive operations, indeed 'at any time' whilst serving on the Western Front. Private William Gregg was 'completely fearless' and 'came through action after action unscathed.' and was 'a fine fighting man . . . we would follow him anywhere', according to the testimony of one of his platoon corporals. What's more, Gregg was the first 'Tommy' to win *three* gallantry awards, and in his case, they ascended in order of merit. His achievements and growing reputation was also reflected in rapid promotions through the ranks during 1917, from Acting Corporal (7 January) and Corporal (2 March) to Acting Sergeant (5 June) and Sergeant (12 December).

Born at Heanor, Derbyshire, at the age of twenty in 1910 William Gregg married a local girl, eighteen-year-old Sarah Hardy, in their parish church on 25 June 1910, his occupation given as 'miner' in the register. He worked at 'Shipley Colliery', part of an old mining complex owned by the Miller-Munday family on their country estate.[36] At the time of William's enlistment into the 13th Battalion, Rifle Brigade (Prince Consort's Own), on 24 November 1914, the Greggs's had a four-year-old daughter.

Six months after volunteering, William was wounded during the Somme offensive. After having endured many weeks in and near the trenches, in 1917 Bill won the first two of his gallantry awards. The Military Medal 'for bravery in the field' was awarded for a courageous act on 4 February 1917. In 'dangerous daylight' he crawled towards a crater containing a body, which he was able to confirm as that of a German soldier. The Distinguished Conduct Medal 'for gallantry in the field', a second-level military decoration (below the Victoria Cross), followed, awarded for actions on 24 November when he carried messages between sections of his battalion whilst under heavy machine-gun fire. Whilst leading a subsequent counter-attack, Gregg was responsible for 'killing and driving off the enemy'.

Aged twenty-eight, Sergeant William Gregg's VC action was on the same Wednesday (8 May) and in the same French village (Bucquoy) as Private William Beesley's. Gregg's extraordinary efforts took place in the afternoon, after the 13th battalion had come under heavy fire from a machine gun post in the Cemetery by the Crucifix, suffering many casualties. He took charge of an assault on the enemy outpost, killing the gun team, capturing the weapon and four men in an adjacent dugout. Rushing another post, he went on to kill two more Germans and captured another. That wasn't, however, the end of his story. Far from it! Responding to a German counter-attack, he led a charge towards another machine-gun post, killing the crew and capturing the gun; but after having to retreat he pressed forward yet again, despite of the obvious danger, holding his position until ordered to withdraw by his commanding officer. Throughout, Gregg was said to have displayed 'the greatest coolness and contempt for danger, walking about encouraging his men and setting a magnificent example'. His confidence may seem to have bordered on self-invincibility but given his past experiences of the dead and dying on war fronts the reality of war must have been apparent to him too.[37]

Much acclaimed as a local hero after the war, William Gregg eventually returned to mining at Shipley colliery until retirement due to ill health in 1959, not long before the pit closed in the 1960s. Earlier, he must have been a tremendous asset to the army in the early years of the Second World War, serving with the Sherwood Foresters until 'age retirement' in 1941. Always faithful to his home community, Gregg passed away on 9 August 1969, aged seventy-nine. Among numerous local commemorations are three that he would have been forever proud: a Sports Hall (as a young man Bill was a promising athlete and amateur footballer), a street named after him, and his medals displayed in the Royal Green Jacket's (Rifles) (RGJ's) Museum at Winchester. William's widow, Sarah, lived to a great age, and was visited by the RGJs on her 100th birthday in 1992, her late husband's array of medals brought to her for 'a final viewing'.[38]

A quadruple gallantry medal hero

William Gregg was one of only eight men to win the VC, DCM and MM in the Great War.[39] He was probably aware of another former miner and war veteran who was a member of this elite club, **John 'Jack' Henry Williams**, a Company Sergeant in the 10th Battalion, South Wales Borderers.[40] From Nantyglo, Monmouthshire, Jack was an experienced blacksmith – initially at Marine and then Cwm collieries – prior to enlisting into the 'Ebbw Vale Borderers' battalion' at the age of twenty-eight on 12 November 1914. It was however his second albeit short spell with the regiment, buying out his discharge in 1906. Quite quickly, Williams was noticed for his leadership skills and this certainly showed on the Somme between 10-11 July 1916 on account of his 'continued and sustained coolness' when handling his men during the capture of Mametz Wood for which he won a senior gallantry award: the DCM. A year later, on 31 July 1917, at Pilckem Ridge (Third Battle of Ypres), his bravery was rewarded again, in the form of an MM. Before year end, another MM – a Bar – was added to his medal tally, for 'selfless action' at Armentières on 7-8 November 1917, when he brought back to the British lines a badly wounded NCO and then returned to safety a colleague found sheltering in a shell hole.

Jack Williams, right arm in a sling, leaving Buckingham Palace after receiving his quartet of medals, including the VC. Daily Mail

It was on the night of 7 October 1918, that Jack Williams won his incredible fourth gallantry award, the most prestigious of all, for his extraordinary actions in battle at Villers Outreaux, France. When his company suffered heavy casualties from a machine gun, Jack ordered a Lewis gun to combat it, rushing the German position and capturing fifteen of the enemy 'single-handedly'. After realizing his solo situation and therefore great vulnerability, several of the prisoners attacked and tried to hold Jack, who repelled their efforts and bayonetted five of them. His extraordinary actions resulting in a final surrender of the cohort. Williams's extreme bravery that night had enabled both his own company and those on the flanks to advance their positions. Shortly afterwards he was struck on the right arm and leg by shrapnel, wounded so badly that he was invalided back to Britain, and declared unfit for service ten days later.

J.H. Williams was one of thirty-two Victoria Cross winners gazetted on 14 December 1918, a good indication of the great activity taking place in the last couple of months before the Armistice was signed. His investiture at Buckingham Palace, on 22 February 1919, was a historic occasion, King George V decorating him with FOUR awards, an unheard-of occurrence at a single ceremony. Still suffering, Jack required emergency medical assistance on the day as his arm wound had opened up, needing attention.

Jack Williams was unfit for a permanent return to mining, but he found employment as a commissionaire at the General Offices of the Ebbw Vale Company, had a spell as a council rent collector, and returned to 'ambassadorial' work in the steel industry until retirement. The most decorated Welsh soldier of the Great War (he also got the Medaille Militaire [France] in 1919), Williams died aged sixty-seven, after admittance to Woolaston Hospital, Newport on 7 March 1953. A legendary local figure, the man himself appears to have regarded his achievements with a mix of pride and modesty, according to family memories.

The twelve other former miners who got VCs in 1918 were: Charles Stone (Royal Field Artillery), Thomas Young (Durham Light Infantry), Albert Mountain (Prince of Wales Own), John Youll (Northumberland Fusiliers), Hugh McIver (Royal Scots), George Prowse (Royal Naval Division), Laurence Calvert (KOYLI), David Hunter (Highland Light Infantry), James Crichton (Aukland Infantry, NZ), William Johnson (Sherwood Foresters), John O'Neil (Prince of Wales Leinster) and Francis Miles (Gloucestershires).

Lives not forgotten

The memories and mementoes of families and the research of local and military historians continue to uncover details of the lives of our recent ancestors who served in the Great War. The final part of this chapter is a sample from more than three hundred of collected 'case studies' of former miners who were on active service in 1918.

Harry's story

> 'My grandfather was a quiet, gentle man who smoked his pipe, drank pints of tea and enjoyed reading westerns . . . He never talked to be about the war, though he did tell my dad the major battles he had been at.' *Gary Wainwright.*[41]

Like my own maternal grandfather (who joined the Army Service Corps as a driver), **Harry Wainwright** was very fond of working with horses. For Harry, a miner's son, 'horse learning' was underground, in charge of a pony as a boy miner, in Allerton Bywater Colliery, a large, deep mine employing well over 2,000 men and boys, situated in the village where he lived, ten miles south-east of Leeds – in the old West Riding of Yorkshire. Probably in the wake of

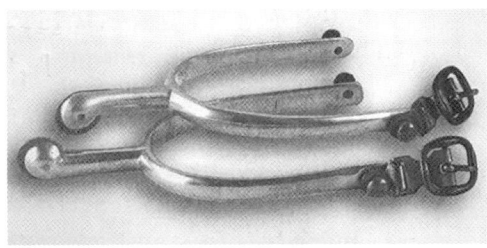

Harry Wainwright's spurs. Gary Wainwright/BBC Books

a rush of local volunteering and with a few pit mates, Harry enlisted at the Leeds recruiting office a few weeks after the start of the war, on 10 October. Service in the artillery, in particular the First West Riding Brigade (and later the 245th Royal Field Artillery (RFA), a Territorial force), certainly gave him plenty of scope for horse-work, albeit in a military context. Standing a little over 5 feet and 4 inches and weighing 10½ stone, he was fairly typical in stature of many other young miner recruits but his 'declared age' of twenty-two (real age only just eighteen) was perhaps an insurance of acceptance.

Harry's role in 'A' Battery meant that he spent the entire war on active service, and in horrific situations on major battlegrounds, including the Somme. He was also at Messines Ridge and in 1918 was involved – under heavy fire and gas attack – in the defensive retreat and then the final barrage of the Allied advance, culminating at 1400 hours on 10 November. The physical and mental impact on his health was enormous. Admitted to Pontefract hospital near his home for almost two weeks at the end of the year, Harry was suffering from an ulcer, according to remnants of his service papers. Surprisingly, he was ordered to return to France after medical discharge and – unlike most other former miners – remained there until February 1919. Demobilisation was delayed further, until March 1920, by which time he had accumulated five and a half years continuous military service.

Harry Wainwright, 'unsung military veteran', married a local miner's daughter, Gertrude Monks, in Kippax parish church on 3 April 1920. Described as 'miner' in the register, he must have been re-employed, probably at his old colliery, Allerton Bywater. Still 'mining' according to his occupational entry in the 1939 Register, Harry lived to a 'good age', passing away at the age of almost eighty in 1976. An emotive reminder of his feelings for 'war horses', and a prized family heirloom, is a pair of spurs that he modified, replacing the sharp pointed stars with centime coins. The majority of former miners who served in the military survived the war and Harry's story, that of an unassuming man who had returned to the pits for a quiet domestic life, may have been typical of at least some of them, their tenancy even overriding the depressing times ahead.

A shell-shocked Sapper

> I look at George's enlistment photograph with a mixture of sadness and also of pride at being his granddaughter and being able to tell his unique story – not all WW1 heroes were killed in action. *Sheila McMillan*[42]

To ensure a quicker homecoming, many miners signed their medical discharge papers as 'fit and well', having no disabilities. This, despite many of them being affected physically and mentally by years of horrendous conditions in and out of the trenches on the Western Front. Some colliery returnees had to accept lighter jobs and less pay than before as they were unable to cope with the physicality of coalface or haulage work. Others kept on working but were unable to continue long term due to the abiding presence of the war on their bodies and minds.

George Bradley aka 'Bull Bradley', a young miner with 'strong hands', worked at Houghton Main, one of the Barnsley area's most productive collieries, situated a couple of miles from Broomhill, near Wombwell, where he lived and was born in 1894. After enlistment – initially with the York and Lancasters – George was promoted to Corporal and transferred to a Special Brigade of the Royal Engineers. His tasks included working at advanced posts, erecting smoke barrages for the protection of advancing troops. It was highly hazardous work, often under artillery fire and onslaughts of gas. On one occasion when preparing a 'gas stunt' a shell landed on George's party, killing his commanding officer, a comrade and seriously injuring several others. The mental impact on him must have been on a lifelong scale. What's more, by the end of service he had been wounded on three occasions.

Corporal George Bradley.
Sheila McMillan

Nevertheless, after demobilisation, Bradley resumed work as a hewer and in the summer of 1921, a 'miners' strike' year, he married Ida Stanger, who was 'in service' at a Barnsley doctor's house. Four children followed, Freda being Sheila McMillan's mother. The 1920s were punctuated by poverty for mining families. In 1926, the miners were 'locked-out' for three months, the Bradleys having to find accommodation in a friend's lodgings, and then a small rented terraced house in Darfield. Houghton Main's 'signing on' records show that George began missing shifts but not due to 'sunshine absenteeism'. A fund-raising social evening was held for the family in December 1927, George now unable to work due to physical and mental disabilities. Fortunately, and exceptionally, their local general practitioner, Dr Castle, himself a veteran of the RFC and a former PoW, referred him to the Star and Garter Home for Disabled Servicemen at Richmond Hill, London. On admission in March 1928 he was found to be suffering from Encephalitis lethargica, a potentially crippling neurological condition that developed from about 1915-16 – alongside 'shell-shock' – which left sufferers with a myriad of ailments including insomnia, chronic anxiety and tremors. The care he got in the home in Richmond was so good that George remaining there for the rest of his life, which sadly ended with a heart attack in 1944. George Bradley's story would have remained largely lost but for the research of his grand-daughter Sheila McMillan who was able to share her family's memories and her research in a little book published in 2023.[43]

A picture palace presentation
The Globe in the south Yorkshire pit village of Conisbrough was packed to the rafters. An 'enthusiastic crowd' on the evening of Saturday 26 October 1918 had assembled – not for silver screen star Mary Pickford – but to see a local war hero. **Quartermaster Sergeant Charles 'Charlie' Chambers Hill** was presented with the Military Medal (MM) for his actions six months earlier, in April 1918, during the retreat from the Somme. As per similar local civic occasions, Hill was also given an Illuminated Address and an inscribed gold watch, courtesy of the Denaby and Cadeby collieries' Heroes Fund.

A Royal Artillery volunteer, Hill had enlisted at the age of nineteen in September 1914, after exiting his clerical job in the offices of Cadeby Main. His MM was awarded 'for great gallantry and coolness during the retreat, when he covered a crossing of the river and kept a field gun in action against the enemy, firing open sights, and only withdrawing the gun (upon receipt of orders) when the German infantry were a few hundred yards away'. Charlie survived the war, though died somewhat prematurely, at the age of fifty-six in 1951. Not to be forgotten, he serves as an example of the many office and ancillary workers at collieries who served and achieved honours in the war, either as volunteers or via conscription.

A most modest signaller
Sapper James 'Jim' Ellis Crooks's war ended on a very high note when in 1918 he won a trio of awards: the MM and Bar and DCM, and was twice mentioned in dispatches. Initially recruited into the West Riding Regiment, he served most of the war in the Signal Service of the Royal Engineers. Usually distinguished by their white and blue armlets, trained (an untrained) signallers were also present in infantry battalions, cavalry regiments and batteries. All were an integral part of front-line operations, laying landlines and relaying messages, often in very hazardous battle conditions. Many signallers gained honours and awards for their bravery. Crawling under fire

Sapper Jim Crooks. Brian Elliott

through No Man's Land to repair phone lines was a not uncommon requirement for these brave men.[44]

Jim Crooks's 'fine war efforts' were marked a few weeks before the end of hostilities by a 'presentation of silverware' at the Free Christian Church in his home village of Bolton-upon-Dearne, near Rotherham. Described as 'quiet and reserved' by an old friend at the occasion, Jim was said to have been 'as cool as a cucumber whenever the enemy was making it hot . . . [especially] after the dark days of March and April [German offensives], when the enemy was knocking at the gates'.[45] All this the more surprising as his transfer into the RE in 1915, appears to have been due to his poor health, as he had previously applied for discharge from the army on the grounds of bronchitis and 'chest problems'.

Acting Corporal Crooks' most notable award, the DCM, was for his actions at Nieppe on 11 April, during the Battle of the Lys, 'for devotion to duty while in charge of a party of linesmen laying a cable' and for completing the task single-handedly – under heavy shell fire – when his men had to drop back to repair breaks in the line. Acknowledging very similar circumstances, his MM was gazetted on 13 September 1918.[46]

Transferred to the Reserve in January 1919, Jim Crooks' marriage in Rotherham in the same year proved to be a somewhat tragic one, his wife Margaret dying prematurely in 1929, the same year as one of their five children; and another son had died three years earlier. Jim Crooks returned to mining, at Hickleton Main, but by the late 1930s was working on the pit top as 'screen hand' – a job still classed as 'heavy work' in the census.[47] Despite all that he had gone through, no doubt with the help of family and mining community, he survived to the age of eighty-two, passing away at Bolton-upon-Dearne in 1974.

The continuing story of a Welsh RAMC veteran

Still in his twenties on discharge from the RAMC in 1919, and 'battle-weary' from the casualty-laden killing grounds of Mesopotamia, **Sergeant Dai Jenkins**[48] was greeted by his wife Rachel on his return to the south Wales village of Kenfig. His family always said that Dai was one of the first to go to war and one of the last to come back. Having Rachel and three small children to support, he had little option but to continue to work as a miner. Such was

Dai Jenkins (1892-1958). Martin Jenkins/BBC Books

The watch given to Dai Jenkins by a grateful Turkish soldier. Martin Jenkins/BBC Books

his interest in helping others, he also continued to serve with the St John's Ambulance Brigade. Dai's medical experience was so notable that he was in demand by both pit mates and local people in need of help, and during the Second World War he served with the Home Guard and as a Special Constable. However, the war continued to effect his own health, bouts of malaria a consequence of his overseas service. His most debilitating ailment however was silicosis, after years of inhaling dust when working underground, a condition that killed him in January 1958, aged sixty-five. One treasured war souvenir passed down the family was a watch given to him by a grateful Turkish soldier who he had nursed back to fitness. Dai's story is one that his family retold – to Paul Atterbury – in a BBC *Antiques Roadshow* programme.[49]

A rehabilitated amputee
Private Percival ('Percy') George Bridges's war came to an abrupt and painful end in a front-line trench in 1918 when shrapnel from a shell burst above him, cutting into his left leg so badly that it required amputation.[50] Hospitalised in Cardiff, Percy was discharged from the army on 3 July 1919, issued with the Silver War Badge, and given a pension in recognition of his disability. Also entitled to the Territorial Force campaign medal, British War Medal and Victory Medal, he had had service in the Somerset Light Infantry (SLI) from January 1916, initially with the 1/4th, and latterly with the 12th battalion. Overseas postings included Mesopotamia and Egypt (1916 to March 1918), prior to his late actions in France and Flanders. Born at Radstock in

Percy Bridges and his Great War medal. Brian Elliott

Somerset in 1895, Percy followed in his father Eli's footsteps as a coalminer, initially working as a 'carting boy', probably at Braysdown Colliery, a small 'family pit' in the pit village of Peasedown St John, where the Bridges had lived from about 1900. Unable to walk without pain or assistance, a return to minework was not a practicable, but Percy's rehabilitation was helped when he learned the craft of 'cobbling'. Married to Florence Button in December 1920, he was able to combine shoe repairing with work as a postman, despite the mobility required in such a rural area. Despite a wooden leg, he cycled to Radstock to collect his mail from the main depot prior to distribution of the same in several villages. By the time of his passing at the age of eighty-two in 1977, he had become a well-known public figure. Percy's son Kenneth Stephen Bridges reported his death and signed the registration form.

Brothers in arms

As we have already seen, miners in and around the coal town of Barnsley rushed in their hundreds to 'join the Colours' at the start of the war. Pals and workmates downed their picks and shovels and brother followed brother to the recruiting office in the Public Hall. Houghton Main was a 'prime' recruitment source, one of the earliest where miners stepped forward in large numbers. On 22 September – only four days after the official start of recruiting for the 13th (Service) battalion of York and Lancaster regiment, nineteen-year-old **Wilfred Cecil Kirk** reacted to the appeal. Keen to follow, less than two weeks later, on 5 October – although underage – was Wilfred's younger brother, seventeen-year-old **Horace Kirk**. To make sure of acceptance, Horace was recorded as the same age as his older brother, and to complicate matters further was allocated soldier number 547, Wilfred logged as 555 earlier. Whatever the circumstances amid the rush, we know that **Samuel 'Sam' Kirk**, a third and the oldest of the brothers – named after his father – joined another local regiment favoured by many miners: the KOYLIs

(King's Own Light Infantry). Sam may have been the first of the family to join the military, though his enlistment papers have not survived in sufficient detail for confirmation.

The brothers had bad wars. Wilfred, part of B Company which was involved in the first wave of the disastrous 'over-the-top' advance at the start of the Somme battle on 1 July 1916, was shot in the chest, dying in hospital in France at the end of the month. Interred at Abberville Communal Cemetery on 2 August,[51] Wilfred's Australian-born widow, Lydia, resident at Cudworth, but living in Barnsley, appears to have received the news only days later, along with 'condolence letters' from the hospital's Sister who had treated him and the regimental Chaplain.[52] Four years married, the couple had had a baby son in 1913, named Horace after his uncle, but the infant died a few days later.

Horace Kirk, a machine gunner, was fortunate to have survived the slaughter that affected so many of his pals on 1 July 1916, held back for an advance the following morning, as their orders were drastically changed in view of the great losses incurred earlier. Horace appears to have done well subsequently, promoted to Acting Lance Corporal on 1 December 1916. However, his mental health deteriorated, wrecked by the horrors seen and experienced. This may have accounted for him – uncharacteristically – absconding over three days in March 1917. Invalided back to Britain for about eight months, he was transferred to the 3rd York and Lancasters as a private, for 'depot duty'. In July 1918 Horace Kirk was given an honourable discharge from the army and granted a pension due to 'shell shock on active service'. It is a credit to his character that he was not tainted as so many were as being weak or even as a coward, and appears to have been treated with some respect by army and medical officers. Living initially at his native residence in Barrow Hill in North Derbyshire, Horace recovered sufficiently to return to Cudworth, Barnsley, and to mine work, his occupation stated as 'coal miner' in the church register when he married his next-door neighbours' daughter, Hilda Quinton, at St Peter's, Barnsley, in 1920. The war, no doubt, was never out of his mind. Horace Kirk died relatively young, in 1948, aged forty-eight.

The senior Kirk brother, Samuel, also survived the war and continued to work as a miner, despite being sent home wounded in 1916, when he was treated in Lord Derby's War Hospital, Warrington. Married to Mary Guest in St John's, Barnsley in 1913, he was still doing 'heavy work' as a hewer by the time of the 1939 Register, passing away at the age of seventy-two in 1964.

The Kirk family had moved from New Whittingham, near Chesterfield to Barnsley in about 1900, part of a wave of migration from north Derbyshire into south Yorkshire. An iron moulder by trade, 29-year-old Samuel Kirk senior and his wife Caroline, aged thirty-one, must have thought their prospects would be far better in the midst of a renown 'coal town'. Two more sons were born, Harry (1901) and Joseph (1903) but little is known about their futures. Samuel himself and his three oldest sons were indeed able to find relatively well-paid work as miners.

Killed in the most tragic of circumstances

Welsh miner **Isaac Roberts**, aged thirty-eight, a former hewer from Diamond Colliery, lost his life when he was working as a medical orderly in the St John Ambulance Brigade Hospital – at Etaples, France – when it was bombed by a German aircraft on the night of 31 May/1 June 1918. Unfortunately, Etaples had several 'war hospitals' as well as barracks, munitions depots and communication lines so was a prime target for German nighttime aircraft raids.

Residing at Ystradgynlais, in the historic southern Welsh county of Brecknockshire, Isaac had volunteered with the RAMC in December 1915 and served on the Western Front as a stretcher-bearer.[53] As an experienced miner, a hewer, he probably had had experience of First Aid. Diamond Colliery, a relatively small concern employing about 300 men and boys, produced the high quality anthracite that was so important for the war effort. Isaac was one of about sixty-five of its workforce known to have served in the Great War.[54] The diary of one RAMC stretcher-bearer, Charles Horner, exposes vividly how this precarious role – usually in a team of four – would have unfolded, unarmed and often under fire amid the terrible mud and blood of battlefields and No Man's Land:

> . . . we lay him on our stretcher with a blanket over him and start our journey back. Carrying in trenches we naturally have to take turns, with one man at each end of the stretcher and the remaining two

Stretcher bearers carry a wounded soldier through the 'Passchendaele' mud. Taylor Library

disposed in front and rear ready to help lifting it, when necessary, head high to negotiate difficult corners. Slings over the shoulders help to relieve the pull on the hands and wrists but still it is hard work . . . we press on toward the road, feeling with our feet what have become regular 'landmarks'.[55]

The context of the above relate to the trenches on the Somme in 1916. Carrying casualties back to British lines and to aid posts also took place over rough, bombed ground and through waterlogged and deep muddy terrain. During the Passchendaele era, progress was both excruciatingly slow and extremely dangerous. No wonder there were about 7,000 stretcher-bearer casualties.

Interred in Etaples Military Cemetery, the poignant words: I HAVE FOUGHT A GOOD FIGHT I HAVE FINISHED MY COURSE were added at the base of his gravestone by his bereaved wife Rachel and his parents.[56] An impressive array of published tributes include an extract from a letter sent by one of Isaac's friends, who said that 'the local hero was mourned by all and was buried by a congregational minister, all the Welsh boys attending [the ceremony]'. The Rt Hon Evelyn Cecil MP, Secretary of the Order of St John, referred to his 'untiring energy and devotion . . . in all his work' and how the organisation was 'indignant at the tragic circumstances in which he met his death by enemy action whilst they were bombing the hospital.' His commanding officer, Lieutenant-Colonel J.E.H. Davies, wrote that 'no man had worked harder or more conscientiously than Private Roberts or showed so much pluck. He was always one to assist others and whilst in the field had gone out to collect wounded regardless of personal danger . . . [and] was an influence for good against others.'[57]

'Keep moving lads'

Thirty-three years after former Lancashire miner **Joseph 'Joe' Rushton** volunteered to serve in the Coldstream Guards he was interviewed and sketched by the artist H.E. Freeth for his series of 'Pit Profiles' in the NCB's *Coal* magazine.[58] Some of his facts may have been a little hazy and like so many he didn't say much about his war experiences but Freeth captured some interesting post-war memories in addition to a wonderful 'likeness' of Joe at a Burnley canalside pit. Leaving school early, Rushton began employment as boy weaver but aged only thirteen was working underground at Hapton Valley Colliery, moving to Habergam for about eighteen months prior to volunteering into the Guards. In the 1911 census, Joseph is recorded as a seventeen-year-old 'pit top labourer', the son of

Joe Rushton, sketched by H.E. Freeth. Freeth family

Lawrence Rushton, a hewer. Whatever the chronology, his medal index card shows his entitlement to 'Victory', 'War' and '15 Star' medals, and that he was in France from 24 August 1915.

Joe missed the Guards' early actions on the Western Front, at Mons and First Ypres but experienced the major encounters of 1916-17, on the Somme, at Passchendaele and Cambrai. Discharged a little early, in March 1918, 'for the mines', he returned to Habergam where he was employed in the very responsible safety role of 'banksman', operating the coal winding gear and shaft signals. His reputation as a senior 'pit top foreman' was enhanced through the 1930s, at Bank Hall Colliery, close to the Leeds-Liverpool Canal, his 'military voice' calling out to the miners at Burnley's biggest pit: 'Keep moving lads! Keep moving'. I imagine few did not argue with Joe's commands, albeit amid a shower of banter.

Shot at Dawn

Walking around Lochnagar Crater back in 2016, one of numerous memorial plaques that I especially noticed and photographed was that of R.W. Simmes (aka Sims, Simes), a Royal Scots soldier who was 'shot at dawn' on 19 May 1918. Serving in the second battalion, **Private Robert 'Bob' William Simmes** was a 28-year-old former miner, originally from Haltwhistle, Northumberland when he was executed 'for desertion'.[59]

Although residing at Brancepath, County Durham, Robert attested at Glencorse, Scotland, on 27 February 1915. His uncle, and later his grandmother were named as 'next of kin'. Of a small stature, 5 feet 5 inches tall, he was shipped to France quite quickly, on 19 May. Simmes's military record shows that he suffered numerous illnesses and injuries throughout his service. On 24 June he was admitted to No. 6 General hospital, Rouen, suffering from diarrhoea, until discharge on 8 July. A few days later he was back in a Rouen hospital due to a hernia, being treated there until the end of the month. Two short spells of treatment at field hospitals followed, on 13 October 1915 and 4 March 1916, both due to myalgia (muscle pain). Then,

'Shot at Dawn' plaque, Lochnagar Crater. Brian Elliott

on 19 March, his immune system probably low, he caught influenza, which require hospitalisation and convalescence. Returned to duty, on 6 May 1916 he absented himself from a morning parade and failed to report for duty until 9.30 in the evening, for which he received fourteen days 'field punishment'. On 14 July, a few days after the start of the Somme battle, Robert was recorded as suffering from shell-shock and got wounded in action near Montauban. Admitted to Etaples hospital, he was again reprimanded for missing a roll call, and given three weeks field punishment after being apprehended by Regimental Police. A further seven days' punishment was issued when he 'feigned' sickness, the start of a prolonged period of indiscipline and punishments until he was injured again, at Serre on 13 November. In hospital for treatment for a bullet wound to his left arm, he was only given a short period of rest prior to a return to duty on the 17th. He was treated for another gunshot wound to his left arm in a field hospital on 11 April 1917 but rejoined his regiment after another short convalescence period. He then lost several days' pay for a 'camp absence' of twelve hours before being admitted to 11th General Hospital, suffering yet again with 'shell-shock'. Almost predicted disciplinary procedures followed when military police apprehended him for absence without leave on 28 September 1917. Discharged from duty on 15 November 1917, Robert was held in custody 'awaiting trial' for 'desertion', 'being in arrest escaping' and 'striking his superior officer being in the execution of his office'.

The 'Shot at Dawn' memorial at the National Memorial Arboretum.
Brian Elliott/Public Domain

Found guilty and executed by firing squad, Private Robert William Simmes was buried in grave III.A.19 in Choques Military Cemetery, France. His three medal entitlements: 1914-15 Star, British War and Victory were forfeited accordingly. It could be reasonably argued that his behaviour was a consequence of the extreme trauma suffered on the Somme over during and after actions, rather than being purposeful and deliberate. On 8 November 2006, a mass pardon was granted to the 306 British Empire soldiers executed for military offences.

Several other former miners appeared to have suffered 'shot at dawn' sentences, including two Sherwood Foresters (Notts and Derby Regiment), for alleged desertion, cowardice and quitting their post, according to recent research: **Corporal Jessie Wilton** (aged forty), 15th Battalion and **Private Arthur Briggs** (twenty-seven), 1st Battalion.[60] All are remembered at the Shot at Dawn Memorial in the National Memorial Arboretum, near Alrewas, Staffordshire.

***

The Armistice – a ceasefire rather than an official end to the war– came into being on 11 November at 11 am, after four bloody years of war.[61] The last British soldier to be killed on this day, at 9.30 in the morning, ninety minutes before the signing, was a forty-year-old York-born coalminer, a hewer, from Leeds, **Private George Edwin Ellison**, shot by a sniper on a woodland patrol near Mons. Ellison was a 'regular' rather than a Kitchener volunteer.[62] George's tragic death was like a bold, giant full stop, as if to say from miners in all the coalfields: 'We did our bit – and a lot more – for the sake of King and Country.'

What was needed next was never delivered. Despite their immense contributions, demobilised miners, their families and communities were not given fair treatment during the 'rebuilding of Britain' after the war.

Chapter Ten

DEMOBBED AND DEMORALISED

Demobilisation was an immense logistical challenge for the Government but achieved with remarkable swiftness. In the year after the Armistice, the size of the British Army was reduced from 3.75 million to 890,000, an astonishing achievement.[1] The process, however, was not without many problems. There were protests and serious unrest in many localities because of the way in which some occupational groups were perceived to have been favoured compared to others.

Coalmining was classed as a priority or 'pivotal' group for release from the Services, as more miners were urgently needed to produce the coal to power post-war reconstruction. Speaking at the end of 1918, the Minister of Reconstruction, Dr Addison[2] said that his department had been very busy making preparations to carry out demobilisation 'at home and abroad' and that arrangements for 100,000 miners to go back to their pits had already been made.[3] This figure, however, was less than half of the total number of miners still in service with the Army, Navy and Royal Air Force. By early January 1919, Sir Guy Calthrop, the Coal Controller, was quoted as saying that 100,000 miners would be released 'in a few days'.[4] Calthrop also mentioned that the men would not be expected to 'go directly from the field of battle to the mines' but given 'an interval of rest before returning underground'.[5]

Troops with 'jobs to go to' were also prioritised, if given a confirmation of this by a previous employer (via a paper 'slip'), so were also released early, adding to a growing toll of ill-feeling from thousands of 'stranded' colleagues and families at home.[6] In response, 'disobedience' broke out on urban streets in January 1919, at such diverse locations as Osterley, Bristol, Kempton Park, Aldershot, Shoreham, Edinburgh, Maidstone and parts of London.[7]

Perhaps not surprising given the rush to get home, the fast-tracking of miners back to Blighty and their pits was subject to abuse. Private Frank Richards, the Royal Welsh Fusilier, considered himself and his mates 'lucky' to be 'sent home first' but noted the following in his memoirs: 'One man admitted to me before he left the Battalion that he had never been on top of a coal pit in his life. I gave him a few particulars and he passed the mining officers' [questions] all right. No doubt he had saved himself another six or twelve months' soldiering. I expect many more did the same.'[8]

Nevertheless, the administrative machinery was so efficient that by the middle of January it was reported widely in the press that 172,927 miners had been released from the Army. Almost 4,000 others were mobilized from service with the Navy and RAF. Regarding re-employment, reports also stated that 'so far, no substantial difficulty had been found in providing work for the men released'.[9] The latter, however, was far removed from what was actually happening at the pitheads.

After Christmas 1918, at Llanbradach Colliery, almost one in three of the workforce were made redundant in order to make way for returnees, generating immediate protests.[10] Action in south Wales was typically swift and dramatic. In January 1919, 6,000 men 'downed tools' from the Dowlais steel and coal complex in support of forty-five displaced men 'engaged at the coal washeries'.[11]

The re-employment of 'miner-soldiers' at collieries thus posed serious problems for the owners and impacted on the existing workforce. It was a situation that dragged on for months. Northumberland was typical, with over a hundred demobbed miners still 'waiting for work' in June 1919.[12] In many instances there was just no capacity at the pits for the returnees to 'resume their old jobs'. Old working faces had been closed and new developments, including the sinking of drifts and new mines had been 'on hold' during the war years. Sinking a new shaft or drift took months, even years to complete. Loyalty and respect for their war service, however, was so strong that in some cases newer recruits to the mines – those employed in wartime to replace miners who had volunteered for military service – were dismissed in order to make room. Where this drastic action happened, it sparked burgeoning discontent from pit to pit and from one coalfield to another.

Clearly, the Government – even the Coal Controller – in their urgent need for production to be boosted, had not thought through the practicalities of re-employment. Strategies to better facilitate this were suggested by both unions and owners. The miners' advocates wanted the working day to be reduced from eight to six hours, which would clearly go against coal-getting targets. The owners broached the setting up of re-training schemes for the returnees, so that they could access employment elsewhere. The Miners' Federation proposed that returning miners without work should be paid 'trade union wages' whilst 'idle'. All coalfield miners' leaders were convinced that the coal industry of Britain would function far better and be far fairer if returned to national ownership.[13]

Trouble at the pits

In the 'gloomy opening' of a 'De-Control' chapter, Sir Richard Redmayne, who had partly taken over the duties of Coal Controller following the death from influenza of Sir Guy Calthrop, had this to say about a most unsettling period for the industry: 'The year 1919 will probably go down in history as the most memorable in the annals of the coal-mining history of the United Kingdom.'[14] Redmayne placed some blame on all the instability on the number of industrial disputes in the country, the 'horizon overcast with great and imminent labour troubles', though he fails to mention the full context of workers' actions, especially in an industry in which he had so much knowledge.[15]

The parlous post-war situation at pitheads was largely a consequence of the industry being in a chaotic state. Returning miners were often unable to get their old jobs back or were in such an unfit physical and/or mental state that they could not do their work in the same way as before. The whole was exacerbated when miners' unions, quite understandably, sought better pay and conditions, though, as Redmayne suggested, it was a time when disputes were on the increase.[16] Indeed, the number of miners 'called out' in 1919 was 906,000, as against 368,000 a year earlier.[17] Strikes were uncomfortably spasmodic for the miner owners but were always started on a point of principle. In Yorkshire, for example, 150,000 miners downed tools for thirteen days in January 1919, concerning a meals dispute involving pit-top workers. In July, a similar number from the same coalfield struck for twenty-nine days, dissatisfied with advances of wages in compensation for a (proposed) reduction in working hours, as recommended in the Sankey Commission report. The Royal Commission on the mining industry, chaired by Justice John Sankey, which included miners' leader Robert Smillie, favoured nationalisation and a seven-hour working day. Sankey's recommendations staved off a national strike but Lloyd George's coalition government ignored the principal recommendations.

In February 1919, matters got more critical for the Government and the mine owners. A miners' strike ballot started, early returns showing a large majority of the coalfield regions in favour of action. Frank Hodge, Secretary of the MFGB, warned the nation of a 'national disaster' if the miners' were not treated fairly.[18] Lloyd George's reaction was to order the deployment of 80,000 special constables to keep peace at the pitheads. It was even reported that machine guns were placed at some, presumably more radical collieries.[19] The withdrawal of national action brought great relief to the Government but left a nasty taste among the miners and their representatives, a condition that simmered and then boiled over in the 'big showdown' two years later. By the end of March, 100,000 miners in various locations 'were idle' for six days in support of better pay.

Back in January, in north Nottinghamshire, following the laying-off of twenty-five miners at Mansfield Colliery 'to make way for returned soldiers', protests spread into Derbyshire: 'Each day saw fresh collieries strike in sympathy . . . At a huge demonstration in Chesterfield, 600-700 miners marched to Sutton in Ashfield and Mansfield . . . It was eventually agreed that the miners would return to work with conditions on both sides.'[20]

The re-employment problem was subsequently tackled with the hurried setting up of a 'conciliation board', which would examine cases where discharged 'soldiers and sailors' were unable to return to their old jobs, despite their willingness to do so. One short-term strategy suggested was a transfer of surplus manpower to a 'less full' colliery nearby.[21] The establishment of local demobilisation committees, were also thought to be a way to deal with the problem, any 'sticking points' referred to a central committee for a final decision. These changes appear to have calmed matters in Nottinghamshire and Wales, at least in the short term.[22]

Amid all the uncertainties at the pitheads, coal – which had doubled in price in the war years – had become even more 'scarce'. In the press, blame was often placed on 'absentee' miners, the 'sunshine pitmen' who missed shifts for inappropriate reasons. Manchester, Bolton and parts of Yorkshire were cited as being 'famous' for 'slackers'.[23] The main cohort of absentees were probably the younger, single miners, with no or little family responsibilities. Miners supporting large families would not, generally, take chunks of time off work for no proper reason. It wasn't so much the court or tribunal fines that put off or got rid of the persistent 'time-offenders' but the pit officials – the deputies, overmen and managers – who had considerable kudos to make any persist absentee at the feel very uncomfortable or facilitate their dismissal; also, the team leaders at the faces would soon exclude men who were regular 'slackers' as their absence affected their own wages. An understandable factor regarding absenteeism, however, was not so much the day off after a heavy night of drinking as the very arduous and dangerous nature of the job, requiring a longer recovery time after a succession of hard shifts. Knowing the hazards and demands of the job, numerous demobilised miners refused – despite the condition of early release – to return to coalmining, a job that still offered 'higher wages' than elsewhere.[24]

A typical post-Armistice 'coal crisis' newspaper report from 1918.
Rotherham Advertiser, 22 November 1918

GREAT SHORTAGE OF COAL.

ECONOMY MUST CONTINUE.

The Coal Controller most emphatically warns the public that in consequence of the cessation of hostilities there must not be any relaxation of the effort to save coal and light and to win coal from the mines. The demands of our Allies (France and Italy) must for some time continue to be of a very exacting character.

The fuel situation in this country is still precarious, and the demand for coal is far from being met. It will be impossible to materially alleviate the coal situation in the immediate future. The public are requested, in the national interest, to continue exercising the utmost economy in the burning of fuel and light at the present time as they so loyally did under a condition of war.

Sick, disabled and 'unfit' miners

No service people came home unaffected by the experiences of war. That is certain. For miners to return to their former jobs, a basic level of fitness was needed, especially regarding the approximate 40% of underground workers classed as 'hewers', the elite group who 'dug the coal' on a 'piece-work' basis. The more coal they got the more pay they got, at least in theory. But, after years away, at the very least, time was needed to adjust, even retrain on the job for a while, a transition that Calthorp, the Coal Controller had sensibly explained. At worst, many of them were just not fit enough to resume work at what were predominantly unmechanised coal faces, places of work most likely subject to accidents such as roof-falls, and occasionally explosions.

On arrival in 'Blighty', miners with serious and life-threatening conditions were treated in 'war hospitals', often located well away from their home areas. New organisations such as the Star and Garter Home in Richmond, Surrey cared for and housed disabled soldiers, helped by the Red Cross. The 900-bed St Mary's Hospital in Roehampton became a leading centre for limb fitting and the rehabilitation of amputees. Its waiting list reached 4,000 in 1918. Other 'war-dedicated' centres of excellence emerged. The Queen's Hospital in Sidcup, founded in 1917 served as a pioneering plastic surgery centre for disfigured soldiers, thanks to the foresight of Harold Gillies.[25] The large, 2,000-bed Beckett Park Hospital at Headingly, Leeds was of national importance because of its expertise in facial surgery and dentistry. New operating procedures and skills enabled some soldiers to have their jaws reconstructed.[26]

The leading advocate for disabled officers and soldiers (and from 1921 president of the newly formed British Legion), Field Marshal Haig, objected to the cruel 'scales of disability' and unfair decisions of the Army and the pension offices.[27] My feature about the 'one-legged' postman Percy Bridges, showed that some former miner amputees managed to find alternative employment in their home areas. In fairness, the sheer scale of the 'rehab' task was an enormous challenge for the authorities. Despite everything, over 360,000 disabled servicemen were found employment in 28,000 firms after training and rehabilitation.[28] Unfortunately, in hard times it was often these employees who lost their jobs first, so by March 1924, 38,000 disabled ex-servicemen remained without work.[29]

The allocation of permanent or temporary 'relief' was based on a central bureaucracy and 2,000 local war pension committees. No wonder decisions and payments were very slow, at least in the first few months of the post-war period. It took pressure from veterans' associations to get the government to make changes, in August 1919, to expedite the process. Even so, the system continued to disadvantage miners, as many suffered from long-term occupational ailments even before enlistment. The Pensions Act stated that claims had to be made within seven years of discharge but of course problems associated with mining often surfaced much later in life. Many of the front-line miners, former tunnellers and others such as RAMC men, suffered from late-onset 'shell-shock', so were usually excluded from financial help, their condition not recorded on file whilst in war service.[30]

Time and understanding really was needed in order for the recently physically and mentally wounded (ailments often combined) to function as they formerly did as underground workers. A so-called 'lighter' job on the pit top was also a demanding prospect for many, and for a family man used to hewer's wages, hard to accept. Walter Benson, a Derbyshire miner 'damaged in body and mind' found it extremely hard to resume his old role, eight months after discharge from hospital. Initially deployed on the pit top and then underground, his demise is empathised well by his great grandson Richard Benson in his tribute book *The Valley*:

> This is dangerous work for the war wounded; not only is there the risk of falling rocks, but veterans' old injuries and mended bones often burst open or break again under the pressure of bending and lifting, and head wounds lead to dizziness and faints in the heat. When injuries have been caused by the work the pit managers will usually pay compensation, but when the war wounded get hurt in the pits the managers say responsibility lies with the Army. The War

'War hospitals'

The new war hospitals for injured servicemen – and the many smaller auxiliary hospitals and convalescence homes – were extremely important in coalfield areas during the First World War. The great rush of miners to enlist, especially via Pals and 'mining regiments', meant that they were subjected to a high combined casualty rate compared with the 'mixed-occupational' make-up of many other regiments and battalions. A detailed study concerning a small Barnsley area colliery village shows that 'killed in action' casualties often emanated from the same or nearby streets and almost half of them came from one pit.[32] The Red Cross database of auxiliary hospitals in the UK records that there were forty-seven in the northern 'colliery counties' of Northumberland and Durham and sixty in the industrial West Riding of Yorkshire. Glamorgan had forty-eight auxiliaries and Monmouthshire eleven. Among the smaller mining counties, Staffordshire had seventeen.[33]

Often converted from mental asylums, war hospitals varied in capacity from about fifty patient places to several thousand, getting more crowded from 1915. Not long after the inmates of the Horton mental asylum in Epsom were discharged (in the early months of 1915) it was filled to capacity with wounded soldiers occupying 24,000 beds, an extraordinary transformation; and was then faced with overflowing wards.[34]

Recovering soldier-patients outside the 'Lancaster' Convalescence Home, Kingstone, Barnsley, in 1917. Brian Elliott

Office often countered that the injuries were not the fault of the mine owner, meaning that the men who fought for 'Britain, the Empire and civilisation' were not entitled to occupational financial support if unable to work again.[31]

Coalmining counties, already noted for their hospitals and convalescence homes for the treatment and rehabilitation of injured miners, also accommodated thousands of war casualties; and were able to expand their provision through the establishment of smaller auxiliary hospitals in converted and adapted buildings. Wharncliffe War Hospital, near Sheffield, had 2,033 beds and by the end of the war had treated more than 37,000 patients.[35]

Numerous convalescence and 'auxiliary hospitals' were also created courtesy of private individuals or organisations. The British Red Cross and the Order of St John of Jerusalem led the way, establishing more than 30,000 in a variety of buildings. Hooton Pagnell, a small country house near Doncaster, was transformed into an 80-bed hospital for wounded soldiers, one of its owners, Julia Warde-Aldham acting as matron. Nearby, Sophia Skipworth of Loversall Hall did a similar job, overseeing a hundred or so wounded inmates at her family's country home. In the coal-town of Barnsley, a vicarage was transformed into a small convalescence home, thanks to the initiative and generosity of a local worthy, Edward G. Lancaster, founder of St Edward's Church at Kingstone. Under the influence of a 'motherly matron', Miss Mary Bellamy, the residents here not only benefited from medical care but had 'social advantages'. A locally-subscribed Driving Fund enabled

Demobbed and Demoralised

inmates to visit local beauty spots and have 'charabanc days' out' to more distant places. By 1919, 562 servicemen had received care and treatment at the Lancaster Convalescence Home.[36] A few miles away, Middlewood Hall, a small country house at the edge of the nearby pit village of Darfield, the home of the Taylor family, accommodated up to thirty-two wounded servicemen and was linked to the aforementioned Wharncliffe War Hospital.[37]

A remarkable example of post-war co-operation between mineworkers began in South Wales. In 1919, workers in the twenty-nine operating pits of the Rhymey Valley combined to pay a weekly subscription of sixpence (2.5p) each, taken from their wages, to raise funds to buy The Beeches, a redbrick mansion in Caerphilly. The building was converted into a hospital for the benefit of men employed in local collieries.[38]

Discharged tunnellers

From late autumn 1918, thousands of war-weary tunnellers, many of whom had endured 1500 days or more of the most extreme form of mining imaginable, were eager to sign their discharge forms, ignoring obvious physical and mental ailments in order to speed the administrative process. Sapper **Alfred Parton**, from Pensnett, near the Black Country town of Dudley, was a veteran of the 177TC. He was gassed but medically discharged in January 1918. His disability pension application failed, leaving a wife and four children without its main breadwinner. Parton's already hard-pressed spouse 'took in washing' to earn 'a few coppers' for food, her husband's future prospects uncertain.[39]

James White, a young Monmouthshire miner, who had attested 'in the field' as a tunneller's mate with the 171TC on 26 August 1915, was admitted to hospital because of 'gas poisoning' on 8 April 1918. It's not known if or when he rejoined his unit. White was discharged from the army after signing the usual 'fit for work' form which included the army's 'cop-out' statement: 'I do not claim to be suffering from a disability due to my military service.' He returned to the UK 'as a miner' on 18 April 1918. However, like most others, he was subsequently, given Class Z certification, which meant that he remained on the Reserve, therefore liable to be called up again in an emergency. White had a wife and a child to support.[40]

A similar story for **James Meek**, a single man who attested as a tunneller's mate with the 171TC about a month after White. Meek was 'gassed' and declared 'a battle casualty' on 9 April 1918 but rejoined his unit on 19 May. Like White, he also signed the 'no disability' form and was despatched home to serve 'as a miner' on 14 December 1918 – but was soon transferred to the Army Reserve, Class Z.[41]

Where detailed studies have been done regarding specific tunnelling companies, much more is known about post-armistice activities and discharges.[42] One hundred and forty-one miners from the 172TC were dispatched 'for priority release' before the end of December 1918.[43] Miners from 253TC were dispatched to Britain a little earlier, 12 December.[44] The extant war diary of the 253TC shows that interviews were actually carried out from 30 November to 5 December for early release of coalminers, 138 released on 10 December and seventeen more later same month. The 254TC had suffered more fatalities of Welsh miners – twenty-four – than in any

Sapper 'Frank' Clarke

Despite the physical and mental battering received by so many of the tunnellers, the fortitude and subsequent survival of some returnees was remarkable. Francis Victor Clarke was an experienced miner, probably at Warsop Vale Colliery, near Mansfield in north Nottinghamshire. Sunk in 1895, at the start of the war it was therefore a relatively new pit, many of the employees housed in terraces built by the Staveley Coal and Iron Company. Aged twenty-six, he volunteering for the Leicestershire Regiment at the start of the war, transferring to the Lincolnshires from 18 February 1915. A year later, his status changed again as he was transferred 'in the field' to the 182TC of the Royal Engineers. Initially, Clarke was involved in creating deep mines and dugouts and this was followed with in digging defensive galleries at Vimy Ridge.

Badly affected by shell-shock, in April 1916 Clarke was hospitalised on three occasions. Almost a year later, he was given twenty-one days field punishment for 'intoxication', but his overall conduct appears to have been good, as he was promoted to Second Corporal in March 1918. Mainly used as 'emergency infantry' the now somewhat scattered 182TC's deployment involved fighting at Bellicourt, Landrecies and Le Cateau. After the Armistice, he was one of the many thousands of former tunnellers dispatched home to work in their former coalmines. However, by the time of the 1939 Register his occupation is recorded as 'chimney sweep'. Despite his war experiences Francis Victor Clarke lived to the age of eighty-two, one of probably many tunneller veterans who were unable to return to mining, at least on a long-term basis.

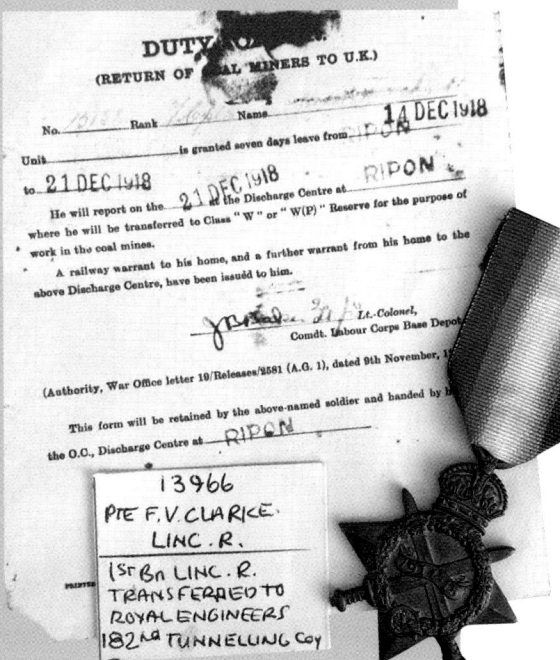

Francis Victor Clarke's 1914-15 Star medal and his 'coal demob' certificate (detail). Brian Elliott

other tunnelling company.[45] McHenry's research shows that the 177TC were kept busy in the Maubeuge area *after* the Armistice, clearing debris from bridges, constructing new bridges and doing salvage work right until the end of January 1919, despite rapidly depleting numbers (123 released for home in last three days of December 1918). By March 1919, a lone officer and nineteen other ranks remained. Badly affected by previous losses at Railway Wood, on 8 December, led by Captain Agner Dalgas, a small party from 177TC returned to the site and re-erected the wooden cross that commemorated the lost tunnellers.[46] In total, the 177TC had sacrificed forty-four men in their actions of the Western Front, mostly miners.[47]

Conscientious objectors (COs) and prisoners of war (PoWs)

Two groups of miners hard for the state to deal with after the Armistice were prisoners of war and conscientious objectors.

The COs who dared to return to their pits faced widespread hostility, and were even excluded via the miners' unions from working alongside existing miners. Forever tarnished as 'shirkers' and 'cowards', they were usually 'cold shouldered' by workmates, treated like 'blackleg' labour. By the spring of 1919 their situation was worsened by politicians making inept and unfair comments about them in parliament. The Marquess of Landsdowne, speaking in the House of Lords on 3 April, warned about conscientious objectors 'taking the jobs of meritorious soldiers'.[48] Earlier, on 26 February in the Commons, Sir William Whitla, a wealthy Unionist, bizarrely suggested that COs 'may be [better] utilised in the cleaning of latrines in different camps at home and abroad'.[49] It took until the summer of 1919 for the last group of COs to be released from the Services into civilian life.[50] The most badly treated of the COs were the so called 'absolutists' who refused to take up arms despite the consequences. They were often incarcerated and after the war disenfranchised by the state and marginalised in society.

About one in twenty of British troops were held as PoWs at the end of the war, most of them (c.165,000) having been captured on the Western Front during August-November 1914 or March-June 1918.[51] Many 'missing' servicemen were also held in captivity but not registered by the Germans, adding to the logistical burden of demobilisation. The poor condition of most PoWs, after years of incarceration and often brutal treatment, meant that their chances of full rehabilitation and a return to mining was limited. Whilst interred, some had undergone 'forced employment' and inhumane conditions in German coalmines, but were too worn out to return to mine work.

Despite the challenges, repatriations of PoWs was largely completed by March 1919. Most former miner-prisoners appear to have been welcomed when they arrived in their home communities.[52] Disability pensions, where issued, however, hardly helped their rehabilitation requirements. Those who managed to return to their pits often had to accept 'lighter jobs' (with less pay).

The condition and prospects of the miner-PoWs, at least 2,000 of them, were worsened in the context of the harsh winter of 1918-19 and the re-emergences of the influenza pandemic. Some, their immune systems understandably low, returned to their families but soon died after catching the 'Spanish lady' (influenza) virus.

Pits, pitmen, production and the 1918-19 pandemic

Preceding the Armistice, during the summer of 1918, the flu virus rushed through industrial towns and districts far faster than in rural areas because so many workers and their families lived in densely packed housing. Coalminers unfortunately were a prime occupational group for the disease as so many of them worked underground in closely confined spaces. Furthermore, as the *The Times* rightly reported, miners were far more likely to suffer more as they often got secondary infections due to existing respiratory problems.[53]

Reports of influenza impacting coalmining regions were therefore common. In Ayrshire, almost one in three of workers at pits in the Galton area were affected, and there was a similar story in Fife, sixty cases reported at Mosside pit, all of them underground employees.[54] Collieries in Northumberland and Durham fared even worse, with up to 70% of their manpower absent because of infection from the deadly virus.[55] It was a similar story elsewhere. In Nottinghamshire, 'a number of employees from the Digby Collieries . . . had to be taken to their homes suffering from the malady';[56] and most of the Nottinghamshire pits were said to be 'struggling' to maintain production, 250 men succumbing to influenza in just one day.[57] A Derbyshire medical officer confirmed the high incidence of the disease among miners, 'who spread the complaint amongst households'.[58] In south Wales, towards the end of June, hundreds of influenza cases were reported, decimating attendance at collieries in Monmouthshire;[59] and in the Rhondda and Merthyr districts miners were badly hit by the spreading epidemic, as were over a thousand mine and steel workers in the Dowlais industrial area.[60]

The epidemic struck at a time when there was already a coal shortage, and when miners were expected to increase production for the 'war effort' and post-war reconstruction. Numerous press reports confirmed that the 'coal crisis' was worsening. In Derbyshire – for example – 'the supply [of coal' was said to be 'very short, due . . . [to] so many miners that are ill with influenza'.[61]

By April 1919, the pandemic was largely over but some 250,000 Britons had died, including many returned miners. Not among the statistics were occasional suicides. Two tragic cases from Barnsley may have been repeated in other coalfield areas. At Worsbrough Dale, Albert Henry Holmes, a 28-year-old miner, had contacted the virus after suffering from pneumonia. In a deranged state, he cut his own throat after his wife had left the house to get him an orange, dying the following day.[62] Less than two weeks later, colliery labourer James William Harper, aged fifty-seven, living at 8 Taylor Row, Barnsley was found drowned in the Dearne and Dove Canal. His cause of death was declared as 'drowned himself during temporary insanity, following influenza' at the inquest.[63]

A country *unfit* for mining heroes

Apart from a short economic 'boom' in 1919-20, unemployment, poverty and labour disputes ravaged coalfield communities post-war, paradoxically at a time when war memorials, rolls of honour, tribute medals and awards were being unveiled and presented.[64] In 1921, the market for the sale of high quality coal abroad had collapsed due to foreign competition, affecting Wales especially, renowned for its anthracite mines.[65] The 'trade slump' of 1920-21 was similar or even more devastating than the great depression of 1930-33.[66]

Miners' unions, however, remained strong in key areas. By February 1921 membership of the South Wales Miners Federation (SWMF aka the FED) exceeded 200,000.[67] And radicalism remained embedded in the Valleys, the formation of the Communist Party of Great Britain in 1920-21 having its roots in the southern Welsh coalfield.[68] As elsewhere, unemployment and wage cuts impacted dramatically on union membership, though the MFGB – after a 'back to the union' campaign – steadied to 148,000 in 1923.[69]

***

Eleven days after the Armistice, David Lloyd George, the former Minister of Munitions (1915-16) and Secretary for War (1916), and now (1918) Prime Minister, arrived in the Black Country town of Wolverhampton and gave what became one of the most iconic, most quoted election speeches in modern political history. The following fifteen words – spoken with typically rousing Welsh passion – resulted in great cheering and applause from a crowd 2,000 crowd jammed into the Grand Theatre:

> What is our task?
> To make Britain a fit country for heroes to live in.

After a few seconds' pause, his remarks were reinforced as follows:

> There are millions of men who will come back.
> Let us make this a land fit for such men to live in.
> There is no time to lose.
> I want us to take advantage of this new spirit.[70]

In the afternoon, Lloyd George was bestowed with a rare honour, the Freedom of Wolverhampton, an award usually given to local worthies and organisations.

Lloyd George 'inspects Barnsley troops' prior to his official reception and award of the Freedom of the Borough. Brian Elliott

Fast-forward three years to 27 August 1921 and the coal town of Barnsley. Lloyd George was given a 'rousing civic welcome' on his much-delayed visit to receive the freedom of the borough.[71] His connection with the town was a tenuous one. Indeed, Barnsley – famous for raising two Pals battalions – received no mention in his huge war memoirs. Fair to say that not everyone was pleased by his visit, which took place in the wake of widespread unemployment, coal rationing and a disastrous lockout of the miners that began on All Fools' Day and lasted three months, to 1 July.

On what became known as Black Friday (15 April 1921), the transport and railway unions – the Triple Alliance – left the miners alone to fight for regional injustices concerning wages and conditions of work. Lloyd George's Coalition government returned the coal industry back into private hands, despite the recommendations of the Sankey Commission and a strong campaign by the TUC for the nationalisation of coal, under the banner slogan 'The Mines of the Nation'.

Miners continued to be paid via complicated 'Price Lists' based on local rather than national agreements. My paternal grandfather got his wages via the sub-contracting or 'butty' system, cash dished out by a lead collier in the most tempting of locations – a pub yard. The whole pay system was open to abuse and riddled with anomalies. In the first quarter of 1921, the average pay of miners was about fifteen to sixteen shillings *per shift*, averaging out at 89s 8d a week, better than most 'industrial wages'. About the time of Lloyd George's visit to Barnsley the *average weekly rate* plummeted to less than 59 shillings, a massive 34 per cent reduction in nine months.[72]

A Labour Party poster issued in 1919 shows a soldier in the trenches, with the same man, bedraggled and unemployed, alongside his wife and baby. Electioneering certainly, but it wasn't far from the truth. Unemployed former soldier-miners wearing ragged trenchcoats were not uncommon sights in and between the coalfield villages and towns of Britain. Former miner-soldiers were reduced to begging for food and attendance at soup kitchens.[73] The depression was so embarrassingly bad that – at the last minute – the Government postponed the 1921 April census.[74]

In Barnsley and elsewhere thousands of mining families faced poverty and near starvation. After weeks on strike there was little food in my grandad's house to feed four small children, other than dwindling vegetables from his allotment. Soup kitchens were provided by the Salvation Army and 'loaf centres' in school playgrounds, church halls and chapel yards, supplied bread to feed the needy. Labour Party MPs managed to get through parliament a Bill that would ensure school children would be fed. The miners' unions did what they could to provide support and relief but most poor families relied on the goodwill of neighbours. The value of the great toil for coal to 'keep the home fires burning' and to power the Army and Navy in the war was a distant memory. This at a time when military medals were still being posted to former service people.

Thus in July 1921, defeated and demoralised, the miners returned to their pitheads glumly aware of false dawns, knowing that the 'subsidisation' of the industry would never happen so long as the economy remained in recession. Welsh coal was back in private hands, any profits from sales in the pockets of the owners whilst miners' pay packets shrunk by a third. Worried at a fall in production, a manager at the Davis and Sons' Ferndale colliery complex, at Rhondda Fach threatened his workers with closure, posting a notice of intent at the pitheads.[75]

Malnourished and angry, in the wake of the 1926 lockout, 270 south Wales miners were the first to march in protest of an unemployment Bill and the injustices of 'relief pay', walking all the way to London in the autumn of 1927, even though the normally supportive Fed, Labour Party and TUC did not back them. The miner-marchers' association with the Communist-linked National Unemployed Workers Movement (NUWM) was disliked by the labour movement.[76] This action was the first of the walking hunger protests, followed in August 1931 by a march to Bristol, when over a hundred of Welsh miners were refused to be seen by the TUC and dispersed by baton-

The poster that said it all: 'Yesterday in the trenches and today unemployed'. Brian Elliott

wielding mounted police. However, the Chartism-like marches of miners from south Wales continued, a somewhat forgotten phase of people protest, over shadowed by the famous Jarrow March of October 1936. Interestingly, and pointedly, 'Jarrow' included a significant contingent of over 500 Welsh miners.

By the mid-1930s, coinciding to George Orwell's visit to Barnsley, one in five of all British miners were unemployed, the same ratio as volunteered to serve in the Great War. Private Turner, a Welsh veteran of the 97th Brigade Machine Gun Corps, spoke up for many others, echoing the voice of Walter Benson quoted above (p.236): 'I would lay puzzling why, why, after all we had gone through in the service of our country, we have to suffer such poverty, willing to work at anything but no work to be had, I only had two Christmases at work between 1919 and 1939.'[77]

In 1921, there were 3,129 coalmines in the hands of about 1,500 separate owners.[78] State control of the coal industry had to wait until well after another world war: Vesting Day, 1 January 1947.

Demobbed and Demoralised • 245

Chapter Eleven

IN MEMORIAM

On a pleasant early December afternoon in 1919, coalminer Horace Holmes – alongside many of his workmates and neighbours – boarded a specially commissioned train at Royston Station bound for Barnsley, almost six miles away. Over 600 passengers disembarked, an aura of chat and banter ensuing on the short walk to the town's Drill Hall. Seated and settled on tea-dressed tables, some 742 people were in attendance, the gathering having expanded with local invitees. The Hall's ambiance of voices hushed with surprising discipline in recognition of the start of a speech from the presiding official, Mr C. Ellison, MD of the large New Monckton Colliery complex. Seated worthies on stage included the Mayor of Barnsley, Alderman (aka Colonel) W.E. Raley and Colonel Sir Joseph Hewitt, the latter instrumental in forming the 13th York and Lancaster's or 'First Barnsley Pals'.[1]

The York and Lancasters' Drill Hall in Barnsley, about the time of the Great War. Paul Wilkinson

The audience listened attentively, especially when, at the start, due respect was paid to the 103 men from Monckton who had 'made the supreme sacrifice in the great war'. The large assembly of returnee soldiers and sailors' 'grand tea' was deemed a mark of appreciation for their service to King and Country. However, the principal purpose of the occasion was to pay tribute – in the form of presentations of inscribed gold pocket watches – to thirty Monckton war veterans who had 'earned decorations' in the war. The most distinguished of these (a Victoria Cross (and Medaille Militaire) holder) was Arthur Shepherd, but among the seven Distinguished Service Medal (DCM) holders present was one of Shepherd's 'collier mates', former Sergeant Horace Edwin Holmes, who was awarded his DCM for his 'coolness and devotion to duty' at the Battle of Cambrai on 20 November 1917.[2] Serving in the West Yorkshire Regiment, the official citation provide a little more detail about his action:

> For conspicuous gallantry and devotion to duty when in charge of the flank platoon of the battalion. When some troops on the flank, being without officers, retired during a counter-attack, after informing his company commander he went across, reorganised and encouraged them, and got back to their line, thus averting a very serious situation.[3]

Sergeant Horace Edwin Holmes DCM (1888-1971)

Born in rural Nottinghamshire, at Weston, Horace was the second of at least five children brought up by his widowed mother Martha. At the age of thirteen, in 1901, he worked as a 'farm yard boy' at Stockwith. The Holmes family had migrated a distance of about forty-five miles, crossing county boundaries, to the booming industrial village of Royston by the time of the 1911 census. Undoubtedly, the pull was the prospect of a better life via employment at the Monckton mining complex which dominated the eastern side of the village. Now aged twenty-three, Horace found work as a 'dataller' or 'day-wage man', his younger brother George, aged twenty, employed as a 'roadman', maintaining the underground routeways. Martha's thirty-year-old daughter Ruth, her husband Christopher (twenty-nine) and their infant son lived at the same address, Christopher doing really well as a hewer at the colliery. Life changed for Horace a year later after he married Nellie Marshall, the daughter of a local miner, in Royston's parish church on 20 May. A salvationist – the village had its own Salvation Army Citadel – Nellie was notable for her work in local public life afterwards.

Horace Holmes in later life, as MP for Hemsworth. Brian Elliott

In Memoriam • **247**

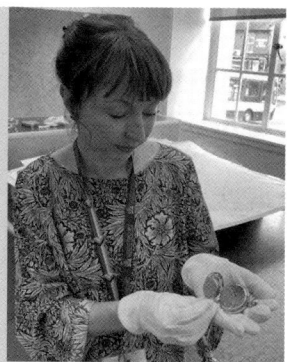

Brian Elliott and Museum assistant examine the gold Holmes pocket watch in Experience Barnsley Archives and Local Studies Library. Brian Elliott

Horace Holmes was an early army volunteer, serving throughout the war with the 2/7th West Yorkshire Regiment (Leeds Rifles), advancing in rank to Corporal and then Sergeant, until his discharge on 19 January 1919. Afterwards he continued working as a miner but also became more and more involved in union activities, rising to become secretary of the Monckton branch of the YMA. Holmes's public profile was enhanced further through service as a magistrate and Labour member of Royston Urban District Council. After the Second World War, in 1946, his local political career advanced significantly when he was elected MP for the large mining constituency of Hemsworth, holding the seat with huge majorities for three successive general elections. His parliamentary career was as a notable one, including spells as PPS to Hugh Gaitskill and Philip Noel-Baker. It was as the Yorkshire Labour Whip that he introduced the young Barnsley MP Roy Mason to the House of Commons on 14 April 1953, when Winston Churchill, almost eighty years old, was in his second premiership. Holmes stepped down from parliament at the 1959 election. Although knighted in recognition of his public services in 1966, he continued to be remembered as a miner and 'war hero', his DCM award marking him out as a distinctive if not unique member of the House of Commons.[4]

The commemorative gold watch presented to Horace Holmes appeared on a well-known online auction site in 2023 but thanks to swift action by Barnsley Museums is now in their care, alongside many exhibits and stories about local people who served in the Great War.[5]

A few weeks later, on 28 December 1919, a Sunday evening, another 'grand gathering' on a similar scale, took place in Barnsley's Public Hall. Described as 'a happy reunion' to honour of 'the glorious record' of the 13th York and Lancaster's Barnsley Pals battalion, the occasion was courtesy of the 13th's former commanding officer, Sir Joseph Hewitt. On entering the Hall, 600 men were presented with briar pipes, tobacco and cigarettes, so there must have been a fair amount of smoke in the air after the assembly had 'done full justice' to what was said to be 'a sumptuous repast ... served by the Barnsley British Co-operative Society. During dinner, entertainment was provided the Wharncliffe Woodmoor Colliery Band and it was also noted that the Hall was 'prettily decorated' for a 'memorable' occasion that afforded ample opportunity for the sharing of 'incidents in camp, trench and No Man's Land' and the 'fitting remembrance' of 'those lads who made the great sacrifice'.[6]

Nearby, in the same week, probably at Dodworth, 'in recognition of the splendid services rendered during the war in the cause of recruiting and looking after the welfare of dependents of the soldier and sailor employees', the manager of Strafford Main and Rob Royd Collieries, Mr Leonard Gill, was presented with 'a gold-mounted umbrella and a smoker's cabinet'.[7]

The end-of-year events described above formed a broad peak of post-war presentations in the Barnsley area that spilled over into the year 1920, celebrations having begun in earnest at several collieries a year or so earlier. The bestowed gifts and presentations had one common feature: to pay due tribute to the 'home and away war heroes', mostly the ordinary miners who contributed so much during the war.

Three post-war presentation events from 1919 – concerning servicemen from Monk Bretton, Barnsley Main and Strafford (Dodworth) collieries – help us to appreciate a trend that was quite common in all coalfield areas. At the February monthly meeting of Monk Bretton Urban District Council, Private Arthur Wing, aged twenty-three, formerly of the 1/5th York and Lancasters, the first cohort (Territorial detachment) of which had mobilised in Barnsley in August 1914, was presented with the Military Medal (MM).[8] A miner at Monk Bretton Colliery, Wing was said to have been 'in the thick of it', throughout his war service. His MM was awarded for bravery in the field on 9 October 1917 on the attack on Passchendaele Ridge when, despite being wounded, he carried a message over 1,200 yards under 'a barrage of fire'. Transferring to the 19th Durham Light Infantry in April 1918, he was discharged from the Army in June 1919.[9]

In recognition of their bravery, 'handsome gold watches' were presented to seven Barnsley Main miners by the Yorkshire Miners Association's junior secretary, Mr S. Roebuck, at the Clarence Hotel, Barnsley on 12 April 1919.[10] A watch was also given to 'Mrs Fitzpatrick', the widow of York and Lancaster soldier Private C. Dodson who lost his life after 'winning the Military Medal'. In due course, all recipients received 'illuminated addresses' as a further recognition of their gallantry. All them had been awarded the MM apart from 22-year-old Sergeant Cyril Fort of the 13th York and Lancasters (1st Barnsley Pals) who had won the DCM 'for conspicuous gallantry and devotion to duty' at Pecq, France, in 1918, the circumstances of his extraordinary action cited as follows:

> On the night of 22 October 1918 . . . he was in charge of a section which was ordered to cross the River L'Escaut, and establish a bridge-head on the eastern side. This necessitated the clearing of a house strongly held by the enemy . . . Working around the flank he rushed the house

and encountered a party of the enemy near the doorway. A short fight ensued, in which several casualties were inflicted on the enemy, he himself accounting for four with his revolver. At this point a number of the enemy . . . rushed out, forcing him to withdraw his men. Though wounded, he personally covered the withdrawal, and was the last to leave. He showed fine courage and did splendid work.[11]

Living close to Barnsley Main Colliery, at Stairfoot, Fort worked as a thirteen-year-old pit-top labourer in 1911, and in 1914 was one of the first teenagers to attest into the 13th York and Lancasters (aged seventeen); and the youngest of the twenty Barnsley Pals decorated with the Distinguished Conduct Medal.[12] Cyril appears to have served as a police officer later, a Sergeant in Sheffield where he passed away at the age of seventy-nine in 1977. His medals, a group of five, were sold at an auction in 2017, for £2,800.[13]

Later on, on the evening of 22 November, a Saturday, in the pit village of Dodworth, a 'memorable gathering' of 300 guests took place at Keresforth Road School, where 'a substantial tea' and 'a musical programme' of entertainment was enjoyed.[14] The occasion was principally for a presentation of two gold medals to two soldiers from Strafford Colliery who had 'gained distinctions' in the war: John Jordan and Lance-Corporal Simpson, both MM holders. From a workforce of about 700, well over 200 had enlisted, twenty-seven of them, about one in eight, losing their lives.

Presentations continued well into the year 1920, at a time, as we have seen, that the slump in the coal industry was biting hard in pit villages and colliery towns. On Saturday evening, 28 February 1920, under the upper-case header 'WOMBWELL HEROES', it was reported that a ceremony was held in the Wesleyan Chapel 'in honour of the men from Wombwell Main Colliery who served in the Great War',[15] presided over by Frank Collingridge, chairman of the Wombwell Main branch of the YMA.[16] The distinguished audience also included the chairman and director of the colliery, Samuel Roberts, MP. Gold watches were presented to six of the employees who had won decorations and 403 gold medals were presented to others who had served during the war, including widows or dependents of the fifty-eight former miners who 'fell in action'. It was said that as many as one in three of the Wombwell Main workforce had volunteered for military service by 1916. Such so-called 'tribute medals' were presented at neighbouring collieries, often carefully inscribed with the names of the recipients.

Gold watches, however, continued to be a popular gift of appreciation presented to former miner employees who had served and/or won distinctions in the war. On the afternoon of 28 March 1920, Palm Sunday, in the Public Hall at Barnsley, eleven 'Wharncliffe Woodmoor heroes' were presented with inscribed gold watches by the Liberal politician and colliery owner Sir William Sutherland who had served as private and press secretary to Lloyd George when the latter was Prime Minister.[17]

Thomas 'Tom' O'Dell's tribute medal

Born and residing at Swinton, O'Dell, worked in his teens as a pony driver at Wath Main Colliery, prior to attesting at Wentworth on 1 October 1915 into the Royal Artillery via the Territorial and Reserve Force (West Riding Royal Horse Artillery). Aged twenty-one and described as a miner in the church register, earlier he married Margaret Lees, on 8 September 1914. The couple's daughter, Kathleen, was born on 18 March 1915, so was only a few months old when his active war service commenced. A 'lead driver', Tom served with the Royal Horse Artillery (attached to the Royal Field Artillery) throughout the remaining four years of the war. He was initially – on 14 September 1918 – transferred to the Army Reserve, and his employer named as the Wath Main Coal Company but was subsequently allocated with a Silver War Badge after appearing a medical board, and confirmed as 'no longer physically fit' on his discharge, 14 December 1918. O'Dell had suffered gas poisoning at some time during his service, a condition that appears to have become more obvious during the latter part of 1918.

Despite his poor health, Tom appears to have been able to collect in person an inscribed bronze tribute medal from the Wath Coal Company, and indeed return to mining at Wath Main long term, as he is recorded as working as a miner in his entry on the 1939 Register, aged forty-six. After retirement he lived to the ripe old age of eighty-nine, and was buried in his home area, Swinton, near Rotherham.

Obverse and reverse views of Tom O'Dell's tribute medal. Brian Elliott

Inscribed gold watches were also presented to 'North Gawber, Woolley and North Gawber [Colliery] heroes' a few weeks later, fifteen former employees awarded them in the Empire Theatre at Darton, the occasion presided over by the YMA president Herbert Smith, with entertainment provide by a male voice choir.[18] Eight of the recipients were ex-'Barnsley Pals', six of these getting MMs and two others (Sergeants) getting Meritorious Service medals.

In Memoriam • 251

Two exceptions were the two men who had been awarded DCMs: Corporal C. Langfield, a KOYLI, and Sergeant E. Dunston, who served with the 32nd Ammunition Column (RFA).

In the Rotherham area, the Wath Main Colliery Company issued 700 inscribed 'welcome home' medals – mostly in bronze but some in gold – in honour of their employees' service in the war. Nearby, at Mitchell Main, Wombwell, the owners and the workmen also presented tribute medals to its 'Great War heroes' – including 'solid gold' specimen for those who had obtained honours. Nearby, 'soldier miners' from Thrybergh Hall and Swinton collieries assembled in Kilnhurst's Church Hall in order to receive presentations provided largely from the donations of their colleagues: 'the men who stayed at home'. It was reported that of the 900 colliery employees, as many as 300 had 'joined the forces' during the war, seven of whom had 'gained distinctions' and two others 'held commissions'. Two hundred of the attendees were given inscribed silver watches and thirty-one accepted the same gifts in honour of a lost relative. Forty more received 'silver medallions with ribands' for their exceptional service, courtesy of a Welcome Home Fund.[19]

The funding of such presentations was facilitated from a variety of sources, from individual acts of generosity to subscriptions and gifts from the miners' unions, colliery management and the coal owners. Also very important were the 'Patriotic' and 'Workmen's Relief' funds', set up mainly to provide

Gold medal presented to 'T. Bruce' of Hickleton Main Colliery, courtesy of the colliery's War Relief Fund. Brian Elliott

Special 'stretcher bearer' medal presented to a Leeds area returning soldier. Brian Elliott

financial help to widows and dependents via regular or one-off provision, but also used to assist the purchase gifts of tribute. The size of some were enormous, in south Yorkshire the Hickleton Main Colliery War Relief Fund accumulated £32,258 from the miners and the colliery company, equivalent to about £1.2 million today.[20]

After the war, 'G. Holmes' who had served in the RAMC as a stretcher-bearer, continued his first aid work at Primrose Hill Colliery at Swillington, near Leeds and was presented with a special medal in honour of his dual duties. The 'coal town' of Ashington, Northumberland, issued 'welcome home' tribute medals for presentation to its returning 'soldiers and sailors'.

Ashington's 'soldiers and sailors' were given welcome home medals. Brian Elliott

Alongside the presentations to individual miner war veterans was the raising of funds for war memorials – also described as cenotaphs – and 'rolls of honour', often placed in churches and chapels, public and private buildings, specially commissioned as tributes to miners. Again, the financing varied from gifts and public subscriptions to joint owners-workers providing the funds. Several fine examples relating to larger collieries appeared during the early 1920s in South Yorkshire – as they did in other mining counties. Unveiled and dedicated on Sunday afternoon, 18 September 1921, 'several thousand people' attended the Manvers Main Memorial, a joint venture between Manvers Main Collieries Limited and its miners, created to honour the 206 former employees 'who had made the supreme sacrifice in the Great War'.[21] Nearby, in 1923, at Thrybergh, near Rotherham, a memorial to more than 300 miners who lost their lives was unveiled, described in chapter 5, above. Near Doncaster, the cenotaph in honour of workmen from Yorkshire Main Colliery commemorating sixty of its former miners was resited in a specially commissioned village memorial garden after the colliery closed and its site obliterated.

Dating from about the same time as the Manvers miners memorial, but subsequently removed to a specially created Memorial Garden, the Cenotaph at Edlington near Doncaster commemorates sixty local miners from Yorkshire Main who lost their lives in the Great War. Brian Elliott

In Memoriam • 253

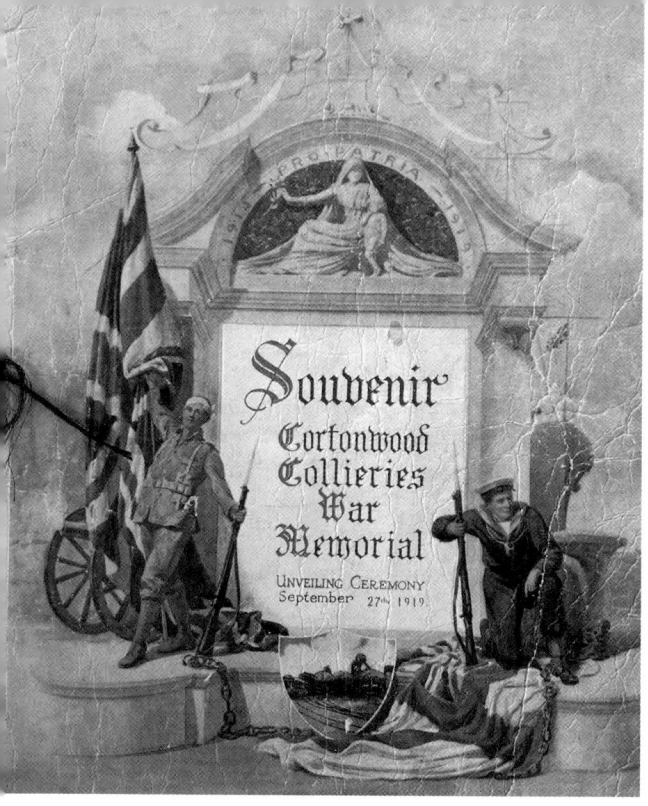

Souvenir brochure published in commemoration of the unveiling and dedication ceremony of the Cortonwood Collieries War Memorial. Brian Elliott

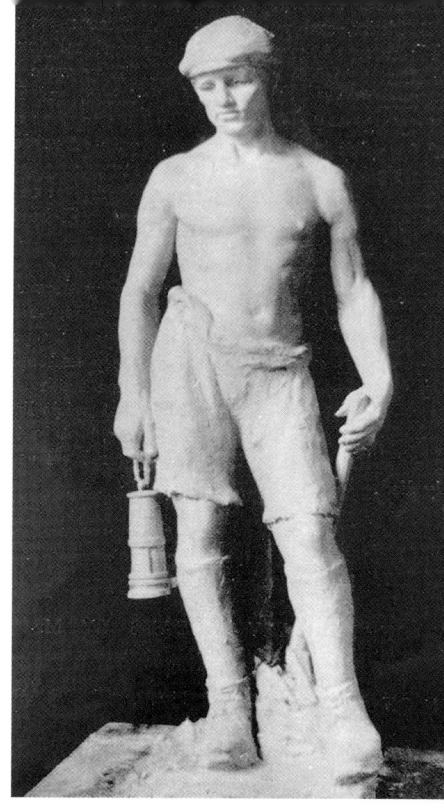

Classical, David-like bronze of a miner, originally placed in one of the two niches of the Cortonwood cenotaph. A 'guarding soldier' completed the figurative composition, the work of a Leicester sculptor, J.H. Morcomb. Brian Elliott

The Cortonwood miners' cenotaph and later pit wheel 'today'. Brian Elliott

One of the most moving Barnsley area miners' cenotaphs and a very early example anywhere, is the one sited in the village of Brampton, near Wombwell. Unveiled and dedicated in front of the colliery offices on 27 September 1919, it's remit was for 'Ninety-four Cortonwood men [who] laid down their lives, and five hundred and seventy-one [who] served in arms', according to a rather grand war memorial souvenir booklet, published by the Mexborough Times (printers) after the ceremony. Bronze figures of a soldier and a miner set in side niches of the 2-metre tall structure greatly enhanced an otherwise fairly plain monument, sculptured by J.H. Morcamb of Leicester. The classical David-like pose of the miner was an especially powerful and emotive symbol. Both figures were subsequently removed as a safeguarding measure, and a commemorative half pit wheel was also added later.[22] Nineteen of the Cortonwood miners who had gained 'military decorations' in the war were given specially-inscribed gold watches.

Thanks to research by a local memorial group, a cenotaph for the 'Dearne towns' of Bolton upon Dearne, Goldthorpe and Highgate was dedicated as recently as 3 August 1914.[23] Although placed for all locals who died in service during 1914-18, such was the scale of mining in the Dearne valley that 180 of the 214 names (84%) were of miners.[24] As one would expect, historic war memorials in coalfield areas usually had a preponderance of miners' names inscribed. The war memorial at Rawmarsh, near Rotherham, a coal and steelmaking area, unveiled fairly late, in 1928, has over 320 'Great War names' inscribed, the vast majority of them former miners. Research has revealed that – as in most other examples – numerous names were missed, as many as sixty absent here.[25]

The Dearne Valley war memorial, Peter Finnegan, Peter Shields and Peter Davies representation the local memorial group in 2014. Dearne Valley Weekender

In Memoriam • 255

Lieutenant Charles Sargeant Jagger MC: war memorial sculptor (1885-1934) [26]

Jagger, the son of Enoch Jagger, a Yorkshire colliery manager from Crigglestone and Mary Elizabeth Sergeant, from Lincolnshire, was a talented artist-sculptor who served with the Artists' Rifles (and then the Worcestershire Regiment) in the First World War. The Jaggers had settled in Wentworth Road, Swinton, the edge of a busy industrial area that included several collieries, Enoch overseeing developmental work, probably at Kilnhurst and its connected pits. Charles's 'practical education' began at the age of fourteen, learning the craft of engraving at the renown Sheffield silversmiths Mappin and Web. Studentships at Sheffield School of Art and the Royal College of Art followed, and then a travelling scholarship that took advantage of short residencies in Rome and Venice. The latter was interrupted in 1914 by military service.

Gassed and wounded several times, in actions at Gallipoli and on the Western Front, his commission with the 4th Worcestershires included the award of a prestigious gallantry medal, the Military Cross, for actions in the Battle of Neuve Eglise during the German Spring Offensive in April 1918, when he suffered serious chest wounds.

Detail of bronze soldier from the Great Western Railway War Memorial, Paddington Station.
Wikimedia Commons/Chris McKenna

After the war, several of the new war memorial committees approached and commissioned Jagger to sculpt memorials for their towns and districts. One the first, in 1918-19, was a large relief, the First Battle of Ypres, for an unexecuted Hall of Fame, is housed at the Imperial War Museum. Another early work, also in relief form, is No Man's Land (1919), now held in Tate Britain. It is an extraordinary tribute to his work ethic that he completed these early works in the wake of his war injury recuperation. Numerous other commissions followed, including sculptures for Manchester and Southsea (1921), Brimington (Chesterfield, in St Michael and All Angel's church) and the Great Western Railway Company (1922), the Royal Artillery (1921-25), Belgium's Anglo-Belgian War Memorial (1922-23), Neeuwpoort (1926-28) and Port Tewfik

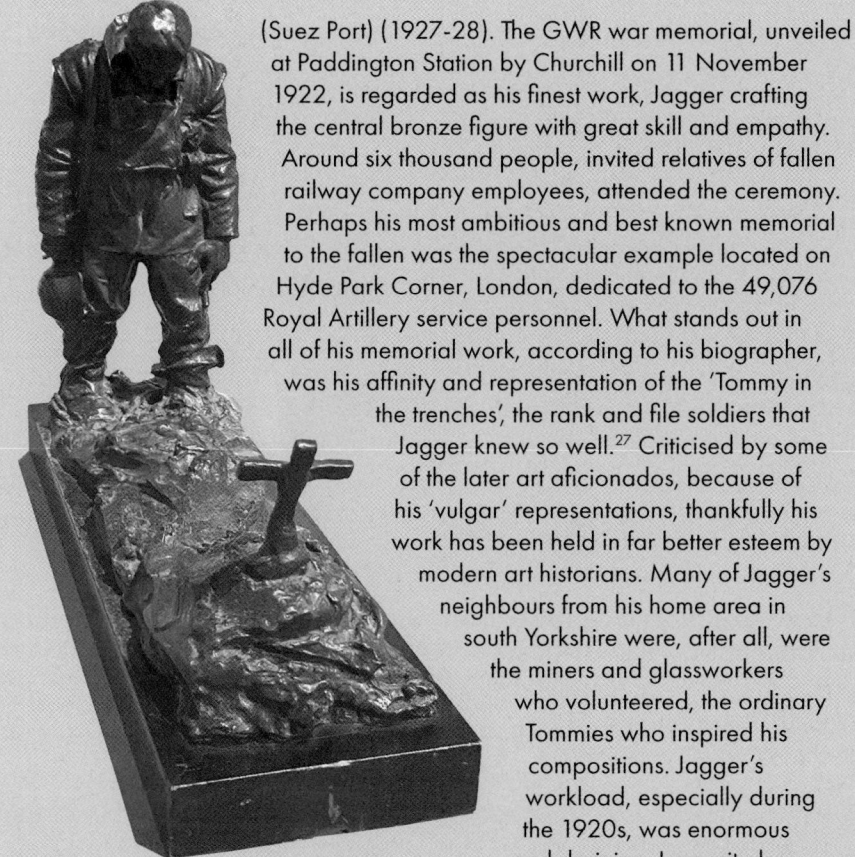

This small bronze composition 'Soldier mourning a friend' is a 1937 reproduction cast by his widow from an original cast and on display at Danum Gallery (Doncaster) museum. Brian Elliott

(Suez Port) (1927-28). The GWR war memorial, unveiled at Paddington Station by Churchill on 11 November 1922, is regarded as his finest work, Jagger crafting the central bronze figure with great skill and empathy. Around six thousand people, invited relatives of fallen railway company employees, attended the ceremony. Perhaps his most ambitious and best known memorial to the fallen was the spectacular example located on Hyde Park Corner, London, dedicated to the 49,076 Royal Artillery service personnel. What stands out in all of his memorial work, according to his biographer, was his affinity and representation of the 'Tommy in the trenches', the rank and file soldiers that Jagger knew so well.[27] Criticised by some of the later art aficionados, because of his 'vulgar' representations, thankfully his work has been held in far better esteem by modern art historians. Many of Jagger's neighbours from his home area in south Yorkshire were, after all, were the miners and glassworkers who volunteered, the ordinary Tommies who inspired his compositions. Jagger's workload, especially during the 1920s, was enormous and draining. Immunity low, Charles Sargeant Jagger, 'colliery manager's son', died unexpectedly, of pneumonia, at his London home, aged only forty-eight years old.

Working alongside Chase Arts for Public Spaces (CHAPS), it was a humbling honour for me to play a part in the unveiling and dedication of the National Miners' Memorial in the National Memorial Arboretum, Alrewas, Staffordshire, in the presence of the Duke and Duchess of Gloucester, on Friday 3 September 2021. This magnificent memorial, created thanks to the hard work of a relatively small group of people, including former miners, has compelling references in its bronze panels to miners at work, including their role as tunnellers in the Great War. The achievement of CHAPS is all the more astonishing given the huge target of raising £100,000 in the context of the Covid-19 pandemic. The public artist responsible for creating the memorial design, Andy DeComyn, was already known for his very moving 'Shot At

The author (centre) with two colleagues, CHAPS members Michael Mellor (left) (Chairperson) and Len Prince (ex-Staffordshire miner), the day before the unveiling and dedication ceremony of the National Miners' Memorial, National Memorial Arboretum, Alrewas, Staffordshire, 2 September 2021. Brian Elliott

'Tunneller': one of twenty-five commemorative bronze plaques forming a frieze around the National Miners' Memorial. Brian Elliott/Andy DeComyn (artist-sculptor)

Dawn' installation in the Arboretum. A specially commissioned publication *A Nation's Tribute* outlines the story of the project and contains my tribute to the miners of Great Britain.[28]

Sited in a peaceful garden enclosure, the National Miners' Memorial provides a place of quiet reflection for visitors to the National Memorial Arboretum, as do the many miners' and coalfield war memorials extant throughout Britain.

'Youth Mourning' by George Clausen (1852-1944). Clausen, an official war artist, painted this moving image in 1916, the horrendous Somme Offensive year. The fiancé of his daughter had been killed in action. A young naked woman is shown in a kneeling and grief-stricken condition set amid a grim, water-filled crater landscape. Public Domain

TIMELINE

1914

Jun 28	Archduke Franz Ferdinand and his wife Sophie assassinated in Sarajevo by a Bosnian Serb.
Jul 28	Austria-Hungary declares war on Serbia.
Aug 1	Germany declares war on Russia.
Aug 3	Germany declares war on France. British mobilize.
Aug 4	Germany declares war on Belgium. Britain declares war on Germany after latter do not cease hostilities.
Aug 7	First British Expeditionary Force (BEF) lands in France.
Aug 11	Lord Kitchener calls for 100,000 new volunteers.
Aug 20	Brussels falls to Germans.
Aug 23	British face German troops at Mons, Belgium
Aug 24	British and French start retreat from Belgium.
Aug 26	Battle of Le Cateau.
Sep 6	First Battle of the Marne. German advance stopped.
Sep 12	First Battle of the Aisne (to 15 Sep).
Oct 19	First Battle of Ypres.
Nov 22	Era of trench warfare begins.
Dec 8	British naval success at Battle of Falklands.
Dec 16	German battleships shell Scarborough and nearby coastal towns of Hartlepool and Whitby.

1915

Jan 19-20	First Zeppelin air raids on Great Britain, at Great Yarmouth and King's Lynn.
Feb	First tunnelling companies of the Royal Engineers formed.
Feb 4	Germans announce start of submarine campaign against merchant shipping in British waters.
Feb 19	British and French attack on the Dardanelles.
Mar 10	British offensive at Neuve Chapelle, France (to 13 Mar).
Apr 22	Second Battle of Ypres (to 25 May). Germans deploy poisonous chlorine gas.
Apr 25	Allied campaign on Gallipoli and Dardanelles.
May 2	German's Gorlice-Tarnow Offensive, Eastern Front (to 18 Sep).
May 7	The liner *Lusitania* sunk by a German submarine.
May 25	Lloyd George is Minister of Munitions in Coalition Government.
May 31	First Zeppelin raid on London.
Jun 30	Flamethrowers used by Germans at Battle of Hooge.
Aug 6	Allied landings at Sulva Bay, Gallipoli.
Aug 21	Last Allied offensive, Gallipoli.
Sep 2	Battle of Loos (to 14 Oct). Six thousand British killed on first day and 60,000 casualties by October.
Sep 25	Poison gas used by British for first time.
Oct 12	Nurse Edith Cavell executed by German firing squad.
Oct 31	Steel helmets issued to British troops.

Nov 24	British Indian army in Mesopotamia retreat to Kt Al Amara after being blocked by Turks at Ctesiphon.
Dec 7	Evacuation of Allied forces from Gallipoli starts (to Dec 20).
Dec 15	Sir Douglas Haig is now Commander-in Chief of the BEF on Western Front.

1916

Jan-July 1	Preparations for Somme offensive on the Western Front, including subterranean offensive/defensive operations.
Feb 21	Battle of Verdun, France.
Apr 29	British-Indian garrison surrenders to Turks at Kt Al Amara, Mesopotamia.
May 31	Battle of Jutland, North Sea
1 Jul-Nov 18	Battle of the Somme, Picardy, France.
	(1 Jul: Battle for Serre)
	(1-13 July: Battle of Albert)
	(14-17 July: Battle of Bazentin Ridge)
	(14 July-15 Sep: Attacks on High Wood)
	(15 July-3 Sept: Battle of Delville Wood)
	(23 July-3 Sept: Battle of Pozieres)
	(3-6 Sept: Battle of Guilemont)
	(15-22 Sept: Battle of Flers-Courcelette: tanks deployed by the British for first time)
	(25-28 Sep: Battle of Morval)
	(26-28 Sep: Battle of Thiepval Ridge)
	(18 Nov: Battle for Serre)
	(13-18 Nov: Battle of the Ancre)
Dec 6	David Lloyd George becomes British Prime Minister.
Dec 18	Battle of Verdun ends.

1917

Feb 1	Germany resumes unrestricted submarine warfare.
Feb 21	German forces withdraw to Hindenburg Line.
Feb 24	Kut retaken by British forces, Mesopotamia.
Mar 11	Capture of Baghdad, Mesopotamia.
Mar 26	First Battle of Gaza, Palestine.
Apr 6	USA declares war on Germany.
Apr 9	Battle of Arras (to 16 May).
Apr 9	Battle of Vimy Ridge, France. (to 12 Apr).
Apr 16	Second Battle of the Aisne, France (to 9 May).
Apr 17	Second Battle of Gaza (to 19 Apr).
Jun 7	Battle of Messines, Belgium (to 14 Jun).
Jul 31	Third Battle of Ypres (Passchendaele), Belgium (to 10 Nov).
Sep 20	Battle of Menin Road Bridge, Belgium (to 26 Sep).
Oct 9	Battle of Poelcapelle, Belgium.
Oct 26	Second Battle of Passchendaele (to 10 Nov).
Oct 31	Third Battle of Gaza (to 7 Nov).
Nov 7	Bolsheviks gain power in Russia.
Nov 20	Battle of Cambrai, France (to 6 Dec).
Dec 11	Liberation of Jerusalem.

1918

Jan 28	Anti-war strikes in German cities.

Feb 24	New Bolshevik government in Russia accepts German peace terms.
Mar 21	Germans launch Spring Offensive – Operation Michael – against British on Western Front.
Apr 1	Royal Flying Corps and Royal Naval Air Service merge to form the Royal Air Force.
Apr 5	Operation Michael ends.
Apr 9	Operation Gneisenau – launched by Germans at the Lys in Flanders.
Apr 29	Germans end their Lys offensive.
May 27	Germans start Operation Blücher on the Aisne. Third Battle of the Aisne (to 6 Jun).
Jun 9	German offensive at Matz is abandoned a few days later.
Jul 15	Second Battle of the Marne.
Jul 16	Bolsheviks murder Tsar Nicholas II and his family.
Aug 8	Battle of Amiens: 'The blackest day of the German army'.
Sep 22	Allies victorious in the Balkans.
Sep 27	Cambrai offensive and Battle of St Quentin, Allies and US troops break through Hindenburg Line.
Oct 4	Germany and Austria request Armistice.
Oct 17	Liberation of Lille and Belgian Channel coast.
Oct 29	Mutiny of German sailors.
Oct 30	Turkey surrenders and signs Armistice with Allies.
Nov 3	Austro-Hungary signs Armistice.
Nov 9	Kaiser Wilhelm abdicates and flees to Netherlands.
Nov 11	Armistice with Germany signed at 5 am and is effective from 11 am.
Nov 21	German navy surrenders.
Dec 14	Lloyd George and Coalition wins large majority in British general election.

1919

Jan	Spanish flu pandemic.
Jan 18	Paris Peace Conference opens.
Feb 14	Establishment of League of Nations.
Jun 21	German High Seas Fleet scuttled at Scapa Flow.
Jun 28	Treaty of Versailles signed and Great War formally ends.
Jul 19	Cenotaph unveiled in London.

1920

11 Nov	Burial of the Unknown Warrior in London and Paris.

ACKNOWLEDGEMENTS

Great thanks are due to my long-time publishers Pen & Sword, in particular to Charles Hewitt for trusting me to tackle a military title, and to Lisa Hooson for her patience and understanding over many years. Jon and Paul Wilkinson (jacket and book designers) have been brilliantly creative as usual and I am grateful to Stephen Chumbley for his copyedit. For social media support thanks are due to Rebecca Carney-Lawther and Rosie Crofts and also to the P&S publicity and marketing team.

Throughout my research, local military historians Jayne Daley and Andrew Featherstone have never tired of my questions and queries though any errors of interpretation remain entirely my own.

From time to time it has been gratifying to contact numerous military historians for permission to quote from their work and/or provide information, and I hope my detailed notes and references, and extensive bibliography will duly appreciate their help and kindness. However, it would be remiss of me not to mention here Paul Reed, Simon Jones, Richard Van Emden, Paul Atterbury and Ritchie Wood. Cyril Pearce was kind enough to read and provide useful comments regarding my chapter about conscientious objectors.

Regarding the many places of research that I have used, I trust that full credit has been given but, again, special thanks are due to staff and facilities at the British Library's newspaper centre at Boston Spa and at the nation's three coalmining museums. The staff at the National Coal Mining Museum's library in particular have been extremely helpful over many visits during the early years of my research, and at the Big Pit curator Ceri Thompson has been a wonderful source of information about Welsh mines and miners. At times I was grateful to be a regular at Rotherham Archives and Local Studies Library, accessing local newspaper and military records housed there. Attending Great War exhibitions nationwide, including several in my native town of Barnsley were perfect backgrounds, indeed inspirational for my work.

Hundreds of case-studies of 'miner-soldiers and sailors' have been used as part of the research and writing of the book, and where possible I have tried to acknowledge the help of descendants such as Jane West (re Enoch Dalton) and Philip Clifford (mine rescue and the Cliffords) accordingly. Indeed, without using previous family and local history research the book would have not, I feel sure, properly connect with general readers.

I am immensely grateful to several author and museum friends who responded to my request to read extracts and summaries about my book, duly providing cover quotes that will remain forever present in my memory. Thanks you so much Paul Brooks, Brian Groom, Dan Jarvis MP, Lynn Dunning, Ceri Thompson, Professor Joann Fletcher, Paul Atterbury, Paul Reed and Ian McMillan.

Last but not least, my heartfelt thanks to Sir Michael Parkinson for writing the Foreword not long before his passing, and also to his son Mike and agent Teresa Rudge in facilitating the process. It was great to meet Michael and correspond with him over several years. He was and is a great inspiration for me to 'keep going' with my writing when the journey gots bumpy.

REFERENCES

Chapter 1: King Coal
1. R.A.S. Redmayne, *The British Coal-Mining Industry During the War* (Oxford Clarendon, 1923), p.2.
2. R. Page Arnot, *The Miners: Years of Struggle* (George Allen & Unwin, 1953). Arnott's figure is taken from 'Parliamentary Report of Departmental Committee to enquire in conditions prevailing in the Coal-mining Industry due to the War (Command Paper 7939)'. Statistics on miner recruitment referred to in subsequent pages extracted from Richard Redmayne's *The British Coal-mining Industry During the War* (Oxford Clarendon, 1923); 'miners' at war' reports in the weekly industrial journal the *Colliery Guardian*; and statistics (usually from the regional miners' unions or MFGB) in local newspapers. Although the majority of miners were union members the actual numbers of miner recruits will usually be greater that those quoted in the press. Also see articles/chapters by Brian Elliott, including 'Their Old Work in Another Guise' (*Memories of Barnsley* [magazine], 29, 2014, pp.6-11) and *A Nation's Tribute* (ed. by Michael Mellor et al), Chapter 2, 'For the Coalminers of the United Kingdom: A Tribute', especially pp.18-21.
3. Statistics relating to employment numbers and production have been largely taken from the annual reports of the mines inspectors (AMIR). For locations of MIRs see Brian Elliott, *Tracing Your Coalmining Ancestors* (Pen & Sword, 2014), pp.55-56 and Chapters 9 & 10 (re national and regional repositories). For this book, the reports have been viewed at the National Coal Mining Museum for England (ncm.org.uk); the annual reports are also reproduced in the *Colliery Guardian* industrial journal (also lodged in the NCMM).
4. Adapted from Roy Church, *The History of the British Coal Industry*, V.4, 1830-1913, p.86.
5. Quoted in the NCB booklet *Pit Pony* (n.d., c.1953); for an excellent account of horses in mines, see Ceri Thompson, *Harnessed* [Colliery Horses in Wales] (National Museum Wales, 2008).
6. Arnot, op.cit., p.24; Brian Elliott, *Images of the Past. Coal Mine Disasters of the Modern Era c.1900-1980* (Pen & Sword, 2017), pp.9-11, 31.
7. Elliott, op.cit., pp.47-48.
8. In a letter to his mother dated January or February 1918, Owen says: 'Wrote a poem on the Colliery Disaster: but I get mixed up with the War at the end. It is short, but oh! Sour!'; Edmund Blunden's notes in *The Poems of Wilfred Owen* (Chatto & Windus [Phoenix Library edition], 1933), pp.98-99, 125.
9. For an authoritative account of industrial disease and ailments associated with mining – within the context of the history of regulation – see Catherine Mills, *Regulating Health and Safety in the British Mining Industries, 1800-1914* (Ashgate, 2010), especially Chapter 1.
10. Sebastian Faulks, *Birdsong* (Hutchinson, 1993); Peter Barton, Peter Doyle and Jonathan Vandewalle, *Beneath Flanders Fields* (Spellmount, 2004).
11. For more on Hackett see Chapter 5.
12. H.F. Bulman and Frederick P. Mills, *Mine Rescue Work and Organization*, 1921, p.9.
13. Extracted and analysed from D.V. Henderson, *Heroic Endeavour* (J.B. Hayward, 1988), pp.57-65.
14. Redmayne, *The British Coal-Mining Industry During the War*, p.258.

Chapter 2: 1914: Pit Duds to Khaki
1. Richard Redmayne, *The British Coal-Mining Industry During the War* (Oxford Clarendon, 1923), p.12.
2. *Barnsley Chronicle*, 19 December 1914.
3. See James Taylor, *The Secret History of the Propaganda Poster* (Saraband, 2013).
4. *Rotherham Advertiser*, 29 May 1915.
5. Consulted in the library of the National Coal Mining Museum for England.
6. *Colliery Guardian*, 14 August 1914.
7. Ibid.
8. *Colliery Guardian*, 16 October 1914.
9. *Western Gazette*, 9 October 1914.
10. 24 October 1914.
11. *Western Mail*, 12 December 1914.
12. *Fife Free Press & Kirkcaldy Guardi*an, 22 August 1914.
13. *Bellshill Speaker*, 25 September 1914.
14. *Durham Chronicle*, 14 August 1914.
15. *Colliery Guardian*, 25 September 1914.
16. *Durham Chronicle*, 23 October 1914.
17. *Yorkshire Evening Post*, 30 November 1914.
18. *Manchester Evening News*, 26 November 1914.
19. *Colliery Guardian*, 24 December 1914.
20. *Nottingham Evening Post*, 28 August 1914.
21. *Colliery Guardian*, 9 October 1914.
22. *Colliery Guardian*, 6 October 1914.

23. *Sheffield Independent*, 10 December 1914.
24. Tim Lynch, *Wakefield in the Great War* (Pen & Sword, 2017), p.35.
25. *Colliery Guardian*, 4 September 1914.
26. *Colliery Guardian*, 25 September 1914.
27. *Colliery Guardian*, 16 October 1914.
28. *Colliery Guardian*, 13 November 1914.
29. *Colliery Guardian*, 25 September 1914.
30. *Colliery Guardian*, 16 October 1914.
31. Scott Lomax, *The Home Front* [Derbyshire in the First World War] (Pen & Sword, 2016), p.60.
32. Lynch, op.cit., p.65.
33. *Colliery Guardian*, 6 November 1914.
34. *Colliery Guardian*, 30 October 1914.
35. *Barnsley Chronicle*, 3 & 24 October 1914.
36. Lomax, op.cit., p.70.
37. *Colliery Guardian*, 24 December 1914.
38. Online sources consulted relating to Streets include warpoets.org.uk [World War 1] and greatwar.co.uk [people]. His book of poetry, *The Undying Splendour*, was published, posthumously, by Erskine McDonald in 1917. See the *Belper News*, 6 July 1917, for a very early obituary-style feature on Streets. His surviving service papers have also been consulted via Ancestry.com.
39. Also see preface to *Undying Splendour*.
40. See, for example, the *Faringdon Advertiser & Vale of White Horse Gazette*, 21 November 1914.
41. Frank Richards, *Old Soldiers Never Die* (Faber, 1933; Parthian [Library of Wales], 2016).
42. Richards, op. cit., pp.45-46.
43. Joseph Murray, Call To Arms (William Kimber, 1980), p.18.
44. *South Yorkshire Times*, 20 September 1914.
45. See museumofmilitarymedicine.org.uk for useful information on the RAMC.
46. Fitch's military record accessed via Ancestry.com.
47. Re Tribute medals and awards see p.250.
48. Brian Elliott, 'Clifford Tyas, the Story of a Barnsley Pal', *Memories of Barnsley* [magazine], 51, Autumn 2019, pp.16-23.
49. R.H. Haigh & P.W. Turner, *Not For Glory* [A Personal History of the 1914-18 War] (Robert Maxwell, 1969).
50. *Rotherham Advertiser*, 10 November 1917.
51. Retrieved via Ancestry.com.
52. Published by Headline, 2005 (see Bibliography). Also see the TV documentaries *Teenage Tommies* (BBC), 2004 and *Secret History: Britain's Boy Soldiers* (Channel 4), 2004.
53. Van Emden, op.cit., p.35.
54. Ibid., p.36.
55. Ibid., p.311.
56. *South Yorkshire Times*, 23 June 1917.
57. Service papers retrieved via Ancestry.com.
58. Retrieved via Ancestry.com.
59. thebeacon-whitehaven.co.uk.

Chapter 3: 1915: Innocence Lost
1. Phil Tomaselli, *Givenchy in the Great War. A Village on the Front Line 1914-1918* (Pen & Sword, 2016), front flap.
2. *Barnsley Chronicle*, 3 July 1915.
3. Robin Barlow, *Wales and World War One* (Gomer, 2014), p.33.
4. Accessed via Ancestry.com.
5. *Rotherham Advertiser*, 3 July 1915; and information from Jayne Daley, 2018.
6. Mick Busby, *They Shall Not Grow Old. Rawmarsh and Parkgate Men Who Fell in the Great War During 1914 and 1915* (n.d., c.2016), pp.114-15.
7. Brian Elliott, 'Their old work in a new guise. Local miners as soldiers in the First World War', *Memories of Barnsley*, 29 (Spring), 2014.
8. *Colliery Guardian*, 22 October 1915; 26 November 1915; 3 December 1915.
9. Ruth Mansbergh, *Whitehaven in the Great War* (Pen & Sword, 2017), p.49.
10. Barry Supple, *The History of the British Coal Industry. Volume 4. 1913-1946* (Clarendon Press Oxford, 1987), p.50.
11. Richard Redmayne, *The British Coal-Mining Industry During the War* (Clarendon Press Oxford, 1923), p.19.
12. *Colliery Guardian*, 1 January 1915.
13. *Colliery Guardian*, 4 January 1915.
14. *Colliery Guardian*, 15 October 1915.
15. Barlow, op.cit., p.8.
16. *Durham Chronicle*, 12 November 1915; J.E. Williams, The Derbyshire Miners (George Allen & Unwin, 1962), pp.530-31.
17. *Colliery Guardian*, 29 January 1915.
18. *Colliery Guardian*, 8 October 1915.
19. *Colliery Guardian*, 17 September 1915.
20. *Colliery Guardian*, 26 March 1915.
21. *Durham Chronicle*, 22 January 1915.
22. *Durham Chronicle*, 3 September 1915.
23. Phil Tomaselli, *Givenchy in the Great War' a Village on the Front Line 1914-1918* (Pen & Sword, 2018), front flap 'blurb'.
24. *Durham Chronicle*, 14 May 1915.
25. *Durham Chronicle*, 11 June 1915, 2 August 1915.
26. Stephen Shannon, *Beyond Praise. The Durham Light Infantrymen who were Awarded the Victoria Cross* (County Durham Books, 1998), pp.15-20.
27. *Colliery Guardian*, 17 March 2016.
28. Shannon, ibid.
29. *Derby Evening Telegraph*, 26 January 1915.
30. *Colliery Guardian*, 22 January 1915.
31. *Colliery Guardian*, 20 August 1915.
32. *Colliery Guardian*, 21 January 1915.

33. *Colliery Guardian*, 12 November 1915.
34. *Colliery Guardian*, 8 January 1915.
35. 10 February 1915.
36. *Colliery Guardian*, 19 March 1915.
37. *Colliery Guardian* & *Durham Chronicle*, 11 June 1915.
38. *Colliery Guardian*, 20 August 1915.
39. *Colliery Guardian*, 8 January 1915.
40. *Colliery Guardian*, 20 August 1915.
41. Scott Lomax, *The Home Front. Derbyshire in the First World War* (Pen & Sword, 2015), p.50.
42. *Colliery Guardian*, 1 January 1915.
43. *Colliery Guardian*, 17 September 1915.
44. *More Dearne Days Remembered*, Dearne Local History Group (booklet), 1988.
45. *Leeds Mercury*, 15 May & 22 July 1915; *Sheffield Evening Telegraph*, 23 July 1915; *Sheffield Independent*, 14 August 1915.
46. *Colliery Guardian*, 2 July 1915.
47. *Colliery Guardian*, 10 December 1915 & 31 December 1915.
48. *South Yorkshire Times* (Retro Supplement), 12 December 2016.
49. Brian Elliott, *Coal Mine Disasters in the Modern Era c.1900-1980* (Pen & Sword, 2017), pp.35-48.
50. Mines inspectors' report accessed in the library of the National Coal Mining Museum for England. Inspectors Ritson, Hay and Humphrys got the MC. The Military Cross (MC) was awarded for acts of gallantry that did not qualify for the VC or Distinguished Service Order (DSO), carried out by junior officers.
51. *Colliery Guardian*, 4 June 1915.
52. Peter Batchelor & Christopher Matson, *VCs of the First World War. The Western Front 1915* (The History Press, 1997 & 2011), pp.169-171; Tomaselli, op.cit., pp.102-105.
53. *Barnsley Chronicle*, 26 June 1915.
54. Jon Cooksey, *Barnsley Pals* (Pen & Sword, 1986), pp.60-63.
55. *Barnsley Chronicle*, 3 July 1915.
56. *Dictionary of National Biography* (entry for Norton-Griffiths).
57. I am indebted to Jane West for letting me have access to her family history research papers.
58. Dave Fordham, *Maltby Main Colliery*(Fed-el-Adoum, 2015), p.20.
59. Simon Jones, *Underground Warfare 1914-1918* (Pen & Sword, 2010), pp.81-82.
60. Robert Graves, *Goodbye to All That* (Jonathan Cape, 1929; Penguin, 1960 [revised ed.], p.102 (Penguin).
61. Graves, op.cit., pp.102-3.
62. Peter Barton et al, *Beneath Flanders Fields. The Tunnellers War 1914-18* (The History Press, 2010), p.147.
63. *Rotherham Advertiser*, 14 August 1915.
64. *Rotherham Advertiser*, 27 November 1915.
65. Edmund Blunden, *Undertones of War* (Cobden-Sanderson, 1928): quoted in 'Poets Memories of Trench War', I Was There, 109, Sir John Hammerton (ed.), 1938.
66. See Chapter 6, pp.127-131.
67. TNA, W095/406/4.
68. *Rotherham Advertiser*, 26 August 1916.
69. Ritchie Wood, *Miners at War 1914-1919* (Helion, 2017), pp.136-39.
70. Mick Manse, *Stories From the Great War* [Vol.1] *The Pooley Miners* (Grosvenor House, 2018), pp.97-102.
71. Barton et al, op.cit., p.65; *Hansard*, 13 December 1922 (ref: 159, 2993-3099).
72. See Bibliography.
73. Barton et al, op cit., pp.148-54.
74. Jones, op.cit.,p.63.
75. Iain McHenry, *Subterranean Sappers. A History of 177 Tunnelling Company RE From 1915 to 1919* (Uniform Press, 2015).
76. McHenry, op.cit., pp.22-23.
77. Ibid, p.26.
78. Ibid, p.27.
79. Ibid, p.38.
80. Site visit by the author on 17 September 2016 courtesy of Leger Tours.
81. Ritchie Wood, *Miners at War 1914-1919: South Wales Miners in the Tunnelling Companies on the Western Front* (Helion, 2017).
82. Wood, op.cit., pp.304-06 (Appendix III).
83. Wood, op.cit., pp.165-67.
84. 16 September 2016, courtesy of St Leger Battlefield Tours.
85. Peter Barton & Simon Jones re the Laboiselle Project: see, for example, BBC News online, 3 November 2011: 'Secrets from Inside a WW1 Trench' (bbc.co.uk) and for LBSG see www.laboiselleproject.com.
86. Shropshire Star.com: 'Soldier's resting place under the Somme found', 2 December 2013.
87. Elaine H. Fisher, *Requiem for Will* (E.H. Fisher, 1997).
88. David Barnes, *Black Mountains* (Dinas Lolfa, 2002).
89. Barnes, op.cit., pp.85-86.
90. Gordon's service papers have not survived, other than his medal index cards, which refer to his enlistment into the 7th Battalion of Prince Albert's (Somerset Light Infantry), and regimental number 14126. His war memoirs were deposited to the People's Collection Wales's free-to-use website peoplescollection.wales.
91. 'Lanchbury' in some reports and records.
92. *Coventry Herald*, 22-23 October 1915.
93. De Ruvigny's Roll of Honour (via Ancestry.com).
94. *Sheffield Daily Telegraph*, 8 October 1915.
95. Graves, op.cit., p.137.

96. TNA: WO 95/2155/2-3.
97. Research by John Morcambe: see *Barnsley Chronicle*, 7 October 1997; and jackclegg.com.
98. Winifred Haward Hotchkiss, *Two Lives* (Yorkshire Arts Circus, 1983), p.15.
99. Ibid.
100. *Barnsley Chronicle*, 23 December 1987.

Chapter 4: Silverwood: A Wartime Colliery
1. Eddie Downes, *Yorkshire Collieries 1947-1994* (Think Pit Publications, 2016), p.520; Alan Hill, *The South Yorkshire Coalfield. A History and Development* (Tempus [The History Press], 2001), p.204.
2. Downes, op.cit., p.519.
3. Brian Elliott, *Images of the Past. Coal Mine Disasters in the Modern Era c.1900-1980* (Pen & Sword, 2017), pp.41-43.
4. Alice Rodgers: online correspondence, 2019; also see Rodgers et al, *God's Coal* (Church of England, 1999).
5. *Rotherham Advertiser*, 10 July 1915.
6. *Rotherham Advertiser*, 15 July 1915.
7. *Rotherham Advertiser*, 13 January 1917.
8. Frank Westwood & Andrew Featherstone, *Rotherham and District War Memorials* (self published, 2003), pp.336-340.
9. *Rotherham Advertiser*, 10 June 1916.
10. *Rotherham Advertiser*, 10 April 1915.
11. *Rotherham Advertiser*, 8 August 1914.
12. *Rotherham Advertiser*, 19 September 1914.
13. *Rotherham Advertiser*, 14 November 1914.
14. *Rotherham Advertiser*, 17 August 1918.
15. *Rotherham Advertiser*, 20 January 1915.
16. *Rotherham Advertiser*, 27 February 1915.
17. *Rotherham Advertiser*, 27 March 1915.
18. *Rotherham Advertiser*, 19 October 1915.
19. *Rotherham Advertiser*, 14 August 1915.
20. *Rotherham Advertiser*, 25 November 1916.
21. Ibid.
22. *Rotherham Advertiser*, 2 September 1916.
23. Ibid.
24. *Rotherham Advertiser*, 21 October 1916.
25. *Rotherham Advertiser*, 7 October 1916 and 4 November 1916.
26. Ibid.
27. *Rotherham Advertiser*, 2 December 1916.
28. *Rotherham Advertiser*, 22 January, 1916.
29. *Rotherham Advertiser*, 19 May 1917.
30. Ibid.
31. *Rotherham Advertiser*, 14 July 1917.
32. *Rotherham Advertiser*, 6 October 1917.
33. *Rotherham Advertiser*, 12 May 1917.
34. *Rotherham Advertiser*, 8 September 1917.
35. *Rotherham Advertiser*, 27 October 1917.
36. *Rotherham Advertiser*, 26 August 1916 and 9 March 1918.
37. *Rotherham Advertiser*, 13 July 1918/ information from Andy Featherstone re the Featherstone family, 2015–20.
38. *Rotherham Advertiser*, 28 July 1917.
39. *Rotherham Advertiser*, 12 January 1918.
40. *Rotherham Advertiser*, 9 March 1918.
41. *Rotherham Advertiser*, 20 July 1918.
42. *Rotherham Advertiser*, 10 August 1918.
43. *Rotherham Advertiser*, 24 August 1918.
44. *Rotherham Advertiser*, 21 September 1918.
45. *Rotherham Advertiser*, 22 & 29 June 1918.
46. *Rotherham Advertiser*, 22 June 1918, 14 and 21 December 1918.
47. *Rotherham Advertiser*, 14 December 1918.
48. *Rotherham Advertiser*, 21 September 1918 and 26 October 1918.
49. *Rotherham Advertiser*, 30 November 1918.
50. *Sheffield Daily Telegraph*, 4 July 1918.
51. *Sheffield Daily Telegraph*, 4 April 1918.
52. *Rotherham Advertiser*, 5 April 1919.
53. *Rotherham Advertiser*, 2 August 1919.
54. *Sheffield Daily Telegraph*, 5 October 1918.
55. *Rotherham Advertiser* & *Sheffield Daily Telegraph*, 2 November 1918.
56. *Rotherham Advertiser*, 13 April 1918.
57. *Rotherham Advertiser*, 26 October 1918.
58. *Dundee Evening Telegraph*, 7 July 1920.
59. For example, *The Scotsman*, 27 September 1921.
60. Family history information courtesy of Leslie Jones, 2015-16.
61. Col. Harold C. Wylly, *York and Lancaster Regiment 1758-1919* (1930) (in Rotherham Archives).

Chapter 5: 1916 Unspeakable Horror
1. For an overview see Roni Wilkinson, *Pals on the Somme 1916* (Pen and Sword, 2006/2014). The late John Biltcliffe's research findings concerning the DLI and the Somme, now digitised and updated by Peter Nelson and Steve Shannon (via durhamatwar.org.uk) provide casualty data (in Part 4) on all battalions.
2. Testimonies based on interviews conducted by my author friends Roni Wilkinson and the late Jon Cooksey in the 1980s: Jon Cooksey, *Barnsley Pals. The 13th and 14th Battalions York & Lancaster Regiment* (Pen and Sword, 1986), pp.70-71, p.209; Roni Wilkinson, 'Barnsley Men and boys on the Somme, 1 July 1916' (*Memories of Barnsley*, 38, Summer 2016), p.13; Lindley also features frequently in Richard Van Emden's 'Boy Soldiers' book, a superb and pioneering piece of the use of First World War oral history: *Boy Soldiers of the Great War* (Headline, 2005).
3. For casualty analysis see Graham Stewart and John Sheen, *Tyneside Scottish* (Pen and Sword, 1999), p.154; see also, generally, John Sheen's *Tyneside Irish* (Pen and Sword, 2010);

for Lochnagar commemorations see online site lochnagar.org (and search: Northumberland Fusiliers Tyneside Scottish 20th-23rd Battalions).

4. Courtesy of Ryton and District War Memorials Project: online searches via rytonwarmemorials.org.

5. rytonwarmemorials.org; memorial halls were built in several County Durham pit villages, Esh Winning Hall (1923) the most notable example: see Norman Emery, *The Coalminers of Durham* (The History Press, 2009), p.179.

6. The LCF has an excellent, informative online site: lochnagarcrater.org.

7. See Paul Atterbury, *Antiques Roadshow World War One in 100 Family Treasures* (BBC Books, 2014), pp.148-151.

8. Stewart and Sheen, p.152; also, generally, Fiona Kay and Neil Storey, *Newcastle Battalions in Action on the Somme* (Tyne Bridge Publications, 2016).

9. By 1915 the Strakers-Love partnership employed more than 3,000 miners in their Brancepeth, Brandon and Willington pits. See dmm.org.uk for details.

10. Brian Elliott, *Images of the Past. Coal Mine Disasters in the Modern Era c.1900-1980* (Pen and Sword, 2017), p.41.

11. Alexander Barrie, *War Underground. The Tunnellers of the Great War* (Tom Donovan, 1961), p.204.

12. Barrie, op. cit., p.220.

13. For a detailed overview of the 171TC see Ritchie Wood, *Miners at War 1914-1919. South Wales Miners in the Tunnelling Companies on the Western Front* (Helion, 2017), especially pp.151-99.

14. Wood, op.cit., p.199.

15. Simon Jones, *Underground Warfare 1914-1918* (Pen and Sword, 2010), pp.149-50; Barrie, op.cit., pp.217-18; Peter Barton, Peter Doyle and John Vandewalle, *Beneath Flanders Fields. The Tunnellers' War 1914-18* (Spellmount, 2004/2010), p.175; also useful is Glo/Coal magazine 'When Dai Met Tommy' issue (Big Pit: Museum of Wales publication, p.50).

16. See, for example, Imperial War Museum's online site iwm.org (Geoffrey Malins/How the Somme was filmed). The silent film was premiered on 10 August 1916 and was a huge success, and shown in all coalfield cinemas.

17. *South Yorkshire Times*, 16 December 1916.

18. *Barnsley Chronicle*, 28 September 1918.

19. For example: Barton, Doyle and Vandewalle, op.cit., pp.290-93; Alan Whitworth, *Yorkshire VCs* (Pen and Sword, 2012), p.118-19; Paul Reed, *Great War Lives* (Pen and Sword, 2010), pp.81-95; Phil Tomaselli, *Givenchy in the Great War* (Pen and Sword, 2016), pp.148-51; Gerald Gliddon, *VCs of the First World War. Cambrai 1917* (The History Press, 2004/12), pp.63-68; among the many online sources see The Comprehensive Guide to the Victoria Cross: vconline.og.uk: and search 'William Hackett'.

20. Jim (J.E.) MacFarlane, *The Bag Muck Strike, 1902-1903* (Doncaster Library Service, 1987).

21. *Rotherham Advertiser*, 2 September 1916.

22. Reports concerning Wilson's death appeared in the *Dalkeith Advertiser*, 12 July 1917 and in the *Mid-Lothian Journal*, 13 July 1917; research on his military service via Arran Sinclair (ajms.medals).

23. Frank Richards, *Old Soldiers Never Die* (Parthian Library of Wales edition, 2016), pp.150-51.

24. Richard Van Emden, *The Somme. The Epic Battle in the Soldiers' Own Words and Photographs* (Pen and Sword, 2016), pp.189, 191.

25. Van Emden, op.cit., p.195.

26. Robert Duncan, *The Mineworkers* (Berlinn Ltd, 2005), pp.146, 190-92.

27. See Terry Newman, *The Hell They Called High Wood: The Somme 1916* (Pen and Sword, 2009); and also Michael Harrison, *High Wood* (Pen and Sword, 2017). Alan Jennings's ww1battlefields.com (Somme/Highwood) is a useful online visitor guide.

28. The online sites findagrave.com and Royal British Legion's everyoneremembers.org has brief memorial and family details relating to Lambert. Wylly's 'York and Lancasters' regimental history (see Bibliography) confirms the Battalion's Somme presence and successful first-day action on 25 September, a Monday, in 1916, despite hostile shelling and 'enemy aeroplanes'.

29. 'Bouzincourt' is the accepted spelling, though 'Bouzencourt' is occasionally used. A blog by Paul Reed ('Somme Battlefields') includes reference to Bouzincourt, accessed via wordpress.com. Also see David Crossland's article 'Whispering Walls' in *Amateur Photographer*, 5 February 2020 (amateurphotographer.com).

30. Bouzincourt Heritage Society, reported by its chairman, Jean-Luc-Rouvillain.

31. According to battlefield archaeologist Gilles Prilaux of the EPCC Somme Heritage Organisation, 3,200 names have been recorded at Nours, 2,200 of them Australian soldiers.

32. Gerald Gliddon, *VCs of the First World War. Somme 1916* (The History Press, 2011), p.215-16.

33. 'Finn' was his family name, though 'Fynn' was used on later records and in reports. W. Alistair Williams's *Heart of the Dragon. The VCs of Wales and the Welsh Regiments, 1914-82* (Bridge Books, 2008), pp.85-91, contains the most comprehensive account.

34. Cwmtillery was a large South Wales colliery complex, consisting of two pits, employing 2,664 men and boys in 1908 and 2,760 in 1918 (see

'Cwmtillery' web page via welshcoalmines.co.uk).
35. *Cornish Guardian*, quoted in W. Alistair Williams, op. cit., p.87.
36. Gliddon, op.cit., (Somme) pp.90-93 (re Davies) and pp.94-97 (re Hill).
37. *Rotherham Advertiser*, 12 August 1916.
38. For example: Richard Van Emden, *Britain's Last Tommies* (Pen and Sword, 2005).
39. livesofthefirstworldwar.iwm.org.uk.
40. Arthur K. Clayton, b.1901;and as featured in Brian Elliott, *Yorkshire Mining Veterans in their Own Words*, (Pen and Sword, 2005).
41. *Barnsley Chronicle* ('Memories' feature), 24 June 2011.
42. Michael Smith and Kay Valentine, *Darfield Remembers the First World War* (Darfield History Society, 2016), p.190 (map) and p194 (stats).
43. Information from Mary Mitchell, Exley's granddaughter, September 2022. The Horton War Hospital functioned from 8 April 1915 (to 17 October 1919), treating more than 44,000 wounded servicemen. The first batch of admissions, were from May 1915: courtesy of information from Laurence Spring, Public Service and Engagement Manager, Surrey History Centre [reference Surrey Heritage/LS/3372/22], 22.9.2022.
44. For musson+relallick and their Light Lines project see mussonrelalick.com: also Katia Harston, 'Somme artwork award', *Barnsley Chronicle*, 16 December 2016.

Chapter 6: 'Conchies': The Men Who Said No
1. See Chapter 4 (pp.52-136) of Adams, op.cit., for examples of Port Talbot miners who were COs. Also telephone conversation with myself (2016) re the '70' figure quoted.
2. Adams (Port Talbot), op.cit., pp.149-53.
3. Ibid, pp.153-59.
4. *Colliery Guardian*, 18 January 1918.
5. *Rotherham Advertiser*, 2 November 1919.
6. *Jarrow Express*, 6 July 1917 & 3 August 1917; *Newcastle Chronicle*, 27 July 1917 & 12 October 1917; also see 'Durham at War' online (durhamatwar.org): Hebburn Lodge.
7. *The Times*, 7 May 1919.
8. *Sunderland Daily Echo*, 10 May 1919.
9. Robert Moore, *Pitmen, Preachers and Politics* (Cambridge UP, 1974), p.200.
10. Part of the Imperial War Museum's Lives of the First World War digital platform as the *Pearce Register of British Conscientious Objectors* (originated by Cyril Pearce). By 2020 the database had grown to almost 20,000 entries; also see reference Pearce in fn.29 (above) and Bibliography.
11. *Colliery Guardian*, 24 March 1916.
12. *Durham Chronicle*, 16 June 1916.
13. *Kilsyth Chronicle*, 26 January 1917.
14. *Fifeshire Advertiser*, 31 August 1918.
15. *Leven Advertiser & Wemyss Gazette*, 23 May 1918.
16. *Burnley Express*, 8 June 1918.
17. *Newcastle Daily Chronicle*, 31 March 1916.
18. Wayne David, *Remaining True: A Biography of Ness Edwards* (Caerphilly Local History Society, 2006).
19. *Oxford Dictionary of National Biography*: Roberts's entry is compiled by Richard Temple.
20. *Wells Journal*, 16 May 1913.
21. ppu.org.uk and menwhosaidno.org (conscientious objectors, 1916-1919): Peace Pledge Union project.
22. Paul Atterbury, *Antiques Roadshow. World War One in 100 Family Treasures* (BBC Books, n.d. [c.2016]), pp.152-55.
23. *London Gazette*, 13 February 1917.
24. Matthew Richardson, *Deeds of Heroes* (Pen & Sword, 2012), p.76.
25. ppu.org.uk.

Chapter 7: 1917: Stinking Trenches and Surreal Landscapes
1. Ernest Hemingway, *For Whom the Bell Tolls* (Jonathan Cape, 1940). Although his book is set in the Spanish Civil War, in 1918 Hemingway was an eighteen-year-old volunteer ambulance driver on the Italian front, until getting seriously wounded by German mortar fire.
2. From the poem 'Memorial Tablet (Great War)'. Written in October 1917, published in Sassoon's 1919 collection *Picture Show*.
3. 'Battlefields Bulletin', June 1916, Leger Tours brochure.
4. *Doncaster Chronicle*, 3 August 1917.
5. Richard van Emden, *The Road to Passchendaele* (Pen & Sword, 2017).
6. Wright & Anderson (eds), *The Victoria Cross and The George Cross* (Methuen, 2013), p.343.
7. *Doncaster Gazette*, 23 February 1917; *Leeds Mercury*, 23 February, 1917.
8. *Doncaster Chronicle*, May 1917.
9. *Litchfield Mercury*, 6 April 1917; *Exeter & Plymouth Gazette*, 29 March 1917.
10. Christine Ball and Nicky Hudson, *The Effects of Rationing on the Home Front 1914-1918* [West Sussex and the Great War Project] (westsussexpast.org.uk); Van Emden & Humphries, *All Quiet on the Home Front* (Headline, 2003).
11. Bill Lawrence, *From Pit Towns to Battlefields* (Mexborough and District Heritage Society, 2015) p.233.
12. Somme exhibition, Experience Barnsley museum, 2016.
13. *Doncaster Chronicle*, 27 April 1917.
14. Andrea Hetherington, *British Widows* (Pen & Sword, 2018); Richard van Emden, *Missing: The*

Need for Closure after the Great War (Pen & Sword, 2020).
15. *Barnsley Chronicle*, 10 November 1917.
16. Also see, for example: Donald Richter, *Chemical Soldiers: British Gas Warfare in World War 1* (Pen & Sword, 2014) and Michael Freemantle, *Gas! Gas! Quick Boys! How Chemistry Changed the First World War* (The History Press, 2014).
17. *Barnsley Chronicle*, 21 April 1917.
18. Michael Smith and Kay Valentine, *Darfield Remembers the First World War* (Darfield History Society, 2016), p.112).
19. *Mexborough & Swinton Times*, 23 June 1917; *Barnsley Chronicle*, 30 June 1917.
20. *Barnsley Chronicle*, 16 September 1916.
21. Ibid.
22. Jon Cooksey, *Barnsley Pals* (Pen & Sword, 2008 edition), pp.204, 219, 227.
23. *Barnsley Chronicle*, 1 December 1917 (and also in a separate piece 'A Royston Pal Killed'.
24. *Colliery Guardian*, 30 March 1917.
25. *Colliery Guardian*, 21 April 1917.
26. *Colliery Guardian*, 22 June 1917.
27. *Colliery Guardian*, 29 June 1917.
28. *Colliery Guardian*, 9 March 1917.
29. *Colliery Guardian*, 18 May 1917.
30. *Barnsley Chronicle*, 27 October 1917.
31. HMMI's general report for 1917, courtesy of NCMME library.
32. Ibid.
33. simonjoneshistorian.com: Myths of Messines.
34. Ritchie Wood, *Miners at War 1914-18* (Helion, 2017), p.193.
35. Wood, op.cit., p.194.
36. Wood, pp.176-77.
37. Wood, p.199.
38. Iain McHenry, *Subterranean Sappers* (Uniform Press, 2015), especially p.224-235.
39. McHenry, op.cit.,p.123.
40. McHenry, p.129.
41. McHenry, p.132.
42. A post by David3497 on Lives of the First World War (IWM) site.
43. McHenry, p.164.
44. Martin Goodman, *Suffer and Survive. Gas attacks, Miners' Canaries, Spacesuits and Bends: The Extreme Life of Dr J.S. Haldane* (Simon & Schuster, 2007), p.304; *The Work of the Royal Engineers in the European War 1914-1919: Military Mining* (book) and G.F.F. Eager, 'The Training of Officers and Men of the Tunnelling Companies of the Royal Engineers in Mine Rescue Work on Active Service in France', paper presented to the Institute of Mining Engineers, 10 Sept 1919 (courtesy of Heroes of Mine: Philip Clifford's site and RE Library Brompton Barracks, Kent).
45. Clifford, 'Heroes of Mine' online site (re: Mines Rescue and the Great War).
46. Simon Jones, *Underground Warfare 1914-1918* (Pen & Sword, 2010), p.182.
47. Ibid.
48. Canaries and resuscitator cages had to be kept at most mines up to the early 1990s, despite the use of modern sensors: *Daily Telegraph*, 18 July 1992.
49. Goodman, *Suffer and Survive*; *The Times* obituary (J.S. Haldane) retrieved from dmm.org.uk (Durham Mining Museum).
50. Jones, op.cit., p.182.
51. Bulman & Mills, *Mine Rescue Work and Organization* (Crosby Lockwood, 1921), pp.51-78.
52. I am grateful to Philip Clifford for his assistance with this information. Also see Philip's 'Heroes of Mine' online site.
53. Brian Elliott, *Images of the Past. Coal Mine Disasters in the Modern Era c.1900-1980* (Pen & Sword, 2017), p.10.
54. *Sheffield Daily Telegraph*, 31 January 1910.
55. Elliott, op.cit., p.23.
56. Ibid., pp.47-48.
57. Gerald Gliddon, *VCs of the First World War: Arras and Messines* (The History Press, 1998/2012), pp.56-60; victoriacrossonline.co.uk; Paul Oldfield, *Victoria Crosses on the Western Front 1917 to Third Ypres 27 January 1917-27 July 1917* (Pen & Sword, 2016), pp.217-20.
58. Elliott, *Coal Mine Disasters in the Modern Era*, pp.75-77.
59. Three more children were born: Mabel (died in infancy), Desmond and Mary.
60. Ceri Thompson (ed) *Glo/Coal* (magazine) (Big Pit/Museum of Wales, n.d.), p.23.
61. Geoffrey Evans, *The Men Who Marched Away. Unlocking the Cynon Valley War Memorials* (Cynon Valley History Society, 2014), p.137 (Appendix II).
62. Stephen Snelling, *VCs of the First World War* (History Press, 1998/2012), pp.216-19; David Bell, *Derbyshire Heroes* (Countryside Books, 2004), pp.9-13.
63. Elliott, *Coal Mine Disasters*, pp.86-87.
64. Wright & Anderson, *The Victoria Cross and the George Cross* (Methuen, 2013), p.523-4;victoriacrossonline.co.uk; *Barnsley Chronicle*: 30 March 1918; 22 February 1919; 29 October 1966; *Yorkshire Post*, 14 February 2008.
65. Mike Bechthold, 'Bloody April Re-Visited: The Royal Flying Corps at the Battle of Arras, 1917' in *British Journal of Military History*, 4, Issue 2, Feb 1918, pp.51-69.
66. *Wigan Observer & District Advertiser*, 10 April 1917.
67. WD P1505276: Col. H.C. Wylly, *York and Lancaster Regiment 1758-1919* (1930, via Rotherham Archives and Local Studies), pp.123-28.
68. Paul Atterbury, *World War One in 100 Family*

Treasures (BBC, 2014), pp.228-29 (and courtesy of Anita Elliott, great granddaughter).
69. *Rotherham Advertiser*, 8 & 15 September 1917.
70. DeRuvigny; *Barnsley Chronicle*, 17 November 1917.
71. Atterbury, op.cit., p.252.
72. The DFM was instituted on 3 June 1918, to non-commissioned officers and men 'for exceptional valour, courage or devotion to duty whilst flying in active operations against the enemy'.
73. *The Gazette*: WW1 Medals (DFM): No.31378 p.7035, supp, 3 June 1919.
74. *Rotherham Advertiser*, 5 July 1919.
75. See Rothesay Stuart Wortley, *Letters of a Flying Officer* (Sutton, 1982);theaerodrome.com (Aces) and wikipedia.org: Ernest Antcliffe.
76. *Rotherham Advertiser*, 5 July 1918.
77. 1939 Register.
78. Shirley Baxter and Hilary McAra, *A Life Too Soon Done. Sheffield General Cemetery and the Great War* (Sheffield General Cemetery Trust, 2016), p.75.
79. *Sheffield Daily Telegraph*, 25 August 1917; *Sheffield Independent*, 27 August 1917.
80. *Colliery Guardian*, 26 January, 1917
81. R.A.S. Redmayne, *The British Coal-Mining Industry During the War* (Oxford, 1923), p.272.

Chapter 8: Behind and Beyond the Wire
1. Robert Jackson, *The Prisoners, 1914-18* (Routledge, 1989).
2. Richard Van Emden, *Prisoners of the Kaiser* (Pen & Sword, 2009); Heather Jones, *Violence Against Prisoners of War in First World War Britain, France and Germany* (Cambridge University Press, 2011); John Lewis-Stempel, *The War Behind the Wire* (Weidenfeld & Nicolson, 2014); David Bilton, *Allied POWS in German Hands 1914-18* (Pen & Sword, 2016); Oliver Wilkinson, *British Prisoners of War in First World War Germany* (Cambridge University Press, 2017).
3. Lewis-Stempel, op.cit., p.xv.
4. Loc.cit.
5. Lewis-Stempel, op.cit., p.144.
6. Ibid., p.109.
7. Ibid., pp.127-8.
8. Quoted in Jackson, p.39.
9. Lewis-Stempel, op.cit., pp.108-9.
10. *Barnsley Chronicle*, 23 February 1918.
11. *South Yorkshire Times*, 4 & 25 January 1919.
12. *Barnsley Chronicle*, 28 December 1918.
13. westernfrontassociation.com.
14. *South Yorkshire Times*, 4 & 5 January 1919.
15. *South Yorkshire Times*, 10 July 1915.
16. *Barnsley Chronicle*, 17 November 1917.
17. *Barnsley Chronicle*, 5 May 1917.
18. Ritchie Wood, *Miners at War 1914-1919* (Helion, 2017); and unpublished research information forwarded by Ritchie to the author about several Welsh tunneller PoWs, August 2021.
19. *South Wales Argus*, 18 June 1915 (courtesy of Ritchie Wood).
20. David Bilton, *Images of War: Allied Prisoners in German Hands 1914-18* (Pen & Sword, 2015), pp.94, 112, 122 & 124.
21. *London Gazette*, Supplement, entry no.6302 (re 24 June 1916 DCM).
22. Lewis-Stempel, op.cit., p.xv.
23. *South Yorkshire Times*, 15 February 1919.
24. The story of Robert Philips formed part of a First World War Exhibition at the Winding House (Elliot Colliery site, Tredegar, Caerphilly), Autumn, 2014; also reported by Latrin Pascoe on Wales online site (20 February 2014) and Ruth Mansfield of *South Wales Argus* (online), 20 August 2014.
25. *The Times*, 22 May 1921.
26. *Cambridge.org*: British Command Paper no.1450 (Supreme Court at Leipzig [judgment], 26 May 1921. Cambridge University Press.
27. Adam Powell, *Soldiering On. British Tommies After the First World War* (The History Press, 2019), p.30.
28. Powell, op.cit., p.37.

Chapter 9: 1918: Dark Pits of War No More
1. Edmund Blunden (ed.), *The Poems of Wilfred Owen* (Chatto & Windus, 1939 edition), p.98.
2. Report of HM Mines Inspectorate (for 1918), summarised in *Barnsley Chronicle*, 10 January 1920; Brian Elliott, *Images of the Past. Coal Mine Disasters of the Modern Era c1900-1980* (Pen & Sword, 2017), p.31; John Benson, *British Coalminers in the Nineteenth Century* (Holmes & Meir, 1980), p.43.
3. Elliott, op.cit., pp.47-48.
4. Kenneth Simcox (2001): retrieved from wilfredowen.org.uk (Wilfred Owen Association).
5. Dominic Hibberd, *Wilfred Owen: A New Biography* (Orion, 2003), p.367.
6. Data from HMMI's annual reports published in the *Colliery Guardian*, housed in the library of National Coal Mining Museum for England.
7. Ibid.
8. Also see *Barnsley Chronicle*, 10 January 1920.
9. *Colliery Guardian*, 10 May 1918.
10. *Colliery Guardian*, 18 January 1918.
11. *Colliery Guardian*, 3 May 1918.
12. *Colliery Guardian*, 19 April 1918.
13. *Colliery Guardian*, 1 March 1918.
14. *Colliery Guardian*, 22 March 1918 & 12 April 1918.
15. *Colliery Guardian*, 12 April 1918.
16. *Colliery Guardian*, 26 April 1918.
17. *Colliery Guardian*, 17 May 1918.

18. *Colliery Guardian*, 18 January 1918.
19. *Colliery Guardian*, 25 January 1918.
20. welshcoalmines.co.uk
21. CWWG: X.C.3A
22. Jasper Conning, 'First World War tunnels to yield their secrets', *Daily Telegraph*, 26 August 2007.
23. Karen Derycke and Lee Inglebrecht, *Memorial Museum Passchendaele 1917 Visitor Guide,* 2015 edition, p.23.
24. Robert Hall, 'Uncovering the secrets of Ypres', 23 February 2007 bbcnewsonline.
25. 'Zonnebeke church dugout' (Wikipedia, accessed 19 May 2022); Derycke and Inglebrecht op.cit., pp.22-25.
26. Register of Tunnelling Company Officers, February 1925.
27. *Colliery Guardian*, 6 December 1918.
28. *London Gazette*, 14 December 1918 (issue 3107, supp.14776); numerous online information about Smith include murtonheritagesociety.co.uk, durhamatwar.org.uk, regimentalmuseumyorkshires.org.uk. Also see 'VC hero . . .' feature in *Northern Echo*, 22 October 2018.
29. *Northern Echo* (cited above).
30. See, for example, Gerald Gliddon, *VCs of the First World War: Road to Victory* (The History Press, 2004, 2014), pp.72-80.
31. *London Gazette*, 18 October 1918 (issue 30967, supp 12488).
32. Quoted by Michael Ashcroft, *Victoria Cross Heroes* (Headline Review, 2006), p.184.
33. CWGC, Beuvry Communal Cemetery Extension, 1.B.7.
34. See, for example, Gliddon, *Spring Offensive*, op.cit., pp.210-18.
35. *London Gazette*, 25 June 1918 (No.30770, pp.7619-20).
36. Now Shipley Country Park, where information is available.
37. *London Gazette*, 25 June 1918 (No.30770, supp.6, p.7619).
38. Gliddon, pp.217-18.
39. Ibid.
40. Gliddon, *Final Days 1918*, pp.131-36; W. Alister Williams, *Heart of a Dragon. The VCs of Wales and Welsh Regiments, 1914-82* (Bridge Books, 2008, pp.284-93).
41. Paul Atterbury, *World War One in 100 Family Treasures* (BBC Books [Ebury Publishing/Woodlands Books], 2014), pp.326-27. Based on *Antiques Roadshow World War 1 Specials*, first broadcast on BBC One.)
42. Sheila McMillan, *Corporal George Bradley and the Star and Garter Home for Disabled Soldiers and Airmen* (The Little Book Club Series [LBS-54], September 2023; also see Sheila's article 'Discovering Grandad' in *Memories of Barnsley*, 69, Spring 2024, pp.26-29.
43. Ibid.
44. royalsignalsmuseum.co.uk.
45. *South Yorkshire Times*, 6 July 1918; 14 September 1918; 21 September 1918; *Leeds Mercury*, 14 September 1918.
46. *London Gazette*, 3 Oct 1918, p.11667 and the Bar on 17 June 1919, p.7645.
47. 1939 Register.
48. See Chapter 7, p.183.
49. Atterbury, op.cit., p.335.
50. Info from research at Radstock Museum, Somerset via Dick Graham, *Peasedown St John and Carlington* (4 vols).
51. findagrave.com: Fn Plot VI.F.8: MemID 56468197.
52. *Barnsley Chronicle*,12 August 1916.
53. *Coal/Glo* (magazine): *When Dai became Tommy* (Big Pit/Museum of Wales).
54. historypoints.org (via Ystradgynlais-history.co.uk).
55. Dale Le Vack (ed.), *Stretcher Bearer! Based on the Diary of Charles H Horton RAMC* (Lion Hudson, 2013).
56. CWGC: Etaples MC, LXV.E.16.
57. yystradgynlaiswargraves.co.uk (Val Trevallion, 2020).
58. Issue No.56, accessed via ncm.org.
59. lochnagarcrater.org (retrieved 5 February 2023).
60. Tim Amos (research) and Dr Denise Amos (writer) via nottsheritagegateway.org.uk (retrieved 28 February 2023).
61. Guns were supposedly silent as from 11am but the war continued at several locations beyond that time, even on the Western Front. The Armistice was for thirty-six days only and then had to be renewed four times, prior to the official end of the war at the (peace) Treaty of Versailles (28 June 1919).
62. His soldier's number (L/12643) confirms this status.

Chapter 10: Demobbed and Demoralised
1. Michael Senior, *The Soldiers' Peace. Demobilizing the British Army 1919* (Pen and Sword, 2018).
2. Christopher Addison, 1st Viscount Addison (1869-1951), previously Minister of Munitions.
3. *Rotherham Advertiser*, 14 December 1918.
4. Calthrop (b.1870), was seconded by the Government from the LNWR. He died in post, a few days afterwards, on 23 February 1919.
5. *Rotherham Advertiser*, 4 January 1919.
6. Adam Powell, *Soldiering On* (The History Press, 2019) p.27.
7. Powell, p.29.

8. Senior, p.38 (from Frank Richards, *Old Soldiers Never Die* [Krijnen and Langley, 2004], p.229.)
9. For example: *Stratford upon Avon Herald*, 24 January 1919.
10. *Colliery Guardian*, 27 December 1918; welshcoalmines.co.uk.
11. *Leeds Mercury*, 28 January 1919.
12. *Newcastle Daily Chronicle*, 4 June 1919.
13. *Montrose Review*, 28 March 1919.
14. Sir Richard Redmayne, *The British Coal-Mining Industry During the War* (Oxford, 1923), p. 215.
15. Ibid.
16. *The Scotsman*, 14 Jan 1919.
17. 'Trade Disputes in GB in 1918 and 1919' in *Monthly Labor Review* [American], Vol 10: no.4, April 1920, pp.227-231. Redmayne quotes slightly different figures: 901,000 in 1919 compared with 356,000 in 1918, 'Coal Industry', p.216.
18. *Evening Mail*, 24 February 1919.
19. Patrick Renshaw, 'Black Friday' in *History Today*, 21, Issue 6, 6 June 1971.
20. *Mansfield Chronicle and Advertiser*, 16 January 191 (quoted in Senior, Soldiers' Peace, p.40).
21. *Colliery Guardian*, 27 December 1918.
22. *Derbyshire Courier*, 8 February 1909; *Leeds Mercury*, 28 January 1919.
23. *Yorkshire Post*, 7 November 1918.
24. *Nottingham Journal*, 9 June 1919.
25. Senior, *Soldiers' Peace*, p.83.
26. John Groom, unpublished notes at Old Moor local history exhibition, RSPB site, Wombwell, Rotherham, accessed 2024.
27. Senior, *Soldiers' Peace*, pp.79-81: There were seven scales of severity, payments scaled accordingly, for example, a fully (100%) disabled private got £27s 6d per week (unskilled labourer rate); the loss of a right arm (90%) meant 24s 9d pw; a left arm lost above the elbow was 16s 6d (60%) and so on. As many as 41,000 men lost at least one limb.
28. Ibid, p.83.
29. Ibid, p.84.
30. PTSD: post-traumatic stress disorder.
31. Richard Benson, *The Valley* (Bloomsbury, 2015 pb edition), p.18.
32. Michael Smith and Kay Valentine, *Darfield Remembers the First World War* (Darfield History Society, 2016), p.190 [map], p.194 [stats].
33. redcross.org.uk (list of auxiliary hospitals in the UK during the First World War).
34. Epsom and Ewell Explorers (eehe.org.uk): Hospital Cluster – Horton.
35. See Dean Hill's and Stuart Reeves's developing site: wharncliffewarhospital.co.uk; also wartimememoriesproject.com.
36. Brian Elliott, *Barnsley 1890s-1990s* (Sutton Publishing, 1999), p.110; Gill Brookes, *Kingstone Remembers the Great War 1914-1918* (Kingstone Parish Heritage Group, 2014), pp.12-18. Also see (re Doncaster, for example) Lynsey Slater and Nicola Fox 'Estate of War' article in *Doncaster Times. At Home, At War* (Issue 1), June 2016, pp.14-16.
37. wharncliffewarhospital.co.uk.
38. *WDYTYA?* Issue 120, Dec 1916, p.62; the former hospital, threatened with demolition was transformed into the Caerphilly Miners' Centre for the Community: caerphillyminerscentre.org.uk.
39. Iain McHenry, *Subterranean Sappers* (Uniform Press, 2015) p.228.
40. Ritchie Wood, *Miners at War 1914-1919* (Helion, 2017), p.197.
41. Wood, *Miners*, p.197.
42. For example, by Wood (*Miners*) and McHenry (*Sappers*).
43. Wood, *Miners*, p.227.
44. Ibid, p.259.
45. Ibid, pp.291-2.
46. Dalgas was responsible for the establishment of the permanent memorial that can be seen on the site today.
47. McHenry, *Sappers*, pp.224-35.
48. Senior, *Soldiers' Peace*, 79: Hansard 3 April 1919.
49. Ibid, p.79.
50. Ibid.
51. Excluding c.20,000 from Dominions: Senior, *Soldiers' Peace*, p.73 [via *Statistics of Military Effort*, TNA].
52. Senior, p.75.
53. Catharine Arnold, *Pandemic 1918* (Michael O'Mara Books, 2018), p.65.
54. *Daily Record*, 14 July 1918.
55. Arnold, *Pandemic*, p.65.
56. *Nottingham Journal*, 2 July 1918.
57. Ibid.
58. *Derbyshire Advertiser*, 6 July 1918.
59. *The Times*, 5 July 1918.
60. *Western Mail*, 27 June 1918.
61. *Yorkshire Post and Leeds Intelligencer*, 8 July 1918.
62. *Barnsley Chronicle*, 22 March 1919.
63. *Barnsley Chronicle*, 5 April 1919.
64. *Soldiering On* (History Press, 2019), Adam Powell's landmark account of British Tommies after the First World War, is a realistic appraisal of the far from smooth transition from military service to civilian life.
65. Robin Barlow, *Wales and World War One* (Gomer, 2014), pp.188-89.
66. Brian Groom, *Northerners. A History* (Harper North, 2022), p.295.
67. Hywell Francis and David Smith, *The FED. A History of the South Wales Miners in the Twentieth Century* (Lawrence and Wishart, 1980), p.28.
68. Ibid, p.28.

69. Ibid, p.31.
70. expressandstar.com [Wolverhampton], 23 November 2018.
71. Brian Elliott 'Census year 1921', *Memories of Barnsley*, 61, Spring issue, 2022.
72. Patrick Renshaw, 'Black Friday', *History Today*, vol.21, Issue 6, June 6 1971.
73. *Coal* (magazine), 'A Land Fit for Heroes' (University of Wales/Big Pit) p.62).
74. Until June.
75. Francis & Smith, p.28.
76. Elin Philips: Chartist 11 eMagazine.
77. *Coal*, p.62.
78. Redmayne, p.270.

Chapter 11: In Memoriam
1. *Barnsley Chronicle*, 13 December 1919.
2. Holmes was presented with his medal before a 'home audience' at the Empire Cinema, Royston: *Barnsley Chronicle*, 12 January 1918.
3. *London Gazette*, 28 March 1918: 267000 H.E. Holmes.
4. The West Yorkshire Archive Service (Wakefield) hold Holmes's papers: C795.
5. Brian Elliott 'Tributes to local miners who served in the Great War' in *Memories of Barnsley*, Issue 70, Summer, 2024.
6. *Barnsley Chronicle*, 3 January 1920.
7. Ibid.
8. The MM was instituted on 25 March 1916, in recognition of 'under fire' acts of bravery not regarded as meriting a DCM [backdated to 1914]. 115,000 were issued in the First World War,
9. *Barnsley Chronicle*, 8 February 1919. Wing had enlisted on 29 May 1914, when still only aged seventeen. In the final year of the war he transferred to the 19th Durham Light Infantry. Working as a general labourer in 1939, he died in Barnsley in 1971, aged 75.
10. *Barnsley Chronicle*, 19 April 1919.
11. Jon Cooksey, *Barnsley Pals* (Pen & Sword, 2008 edition), p.309.
12. Ibid., p.308.
13. By Noonans of Mayfair.
14. *Barnsley Chronicle*, 29 November 1919.
15. *Barnsley Chronicle*, 6 March 1920.
16. Collingridge, a former local miner, was elected MP for Barnsley in 1938, serving until his death in 1951.
17. *Barnsley Chronicle*, 3 April 1920.
18. *Barnsley Chronicle*, 5 June 1920.
19. *South Yorkshire Times*, 3 January 1920.
20. Bank of England: Inflation Converter (online).
21. *South Yorkshire Times*, 24 September 1921.
22. Brian Elliott, *Tracing Your Coalmining Ancestors* (Pen and Sword, 2014), p.128.
23. *Dearne Valley Weekender*, 31 July 2014.
24. Information courtesy of Peter Davies, 4 August 2014.
25. Elliott, op cit., p.129.
26. For information about Jagger see, for example, *The (Sheffield) Star* newspaper's 'Retro' supplement: 'How artist carved a career sculpturing war memorials' by Peter Tuffrey.
27. Ann Compton, *The Sculpture of Charles Sargeant Jagger* (2004, Ashgate Publishing).
28. Michael Mellor, Patricia Mellor, Len Prince and Jacqueline Prince (eds), *A Nation's Tribute* (Chase Arts for Public Spaces [CHAPS], 2021).

BIBLIOGRAPHY

Adcock, A. St. John, *For Remembrance. Soldier Poets who have fallen in the Great War* (Hodder & Stoughton, 1920).
Adams, Philip, *Not in Our Time. War Dissent in a Welsh Town* ([Briton Ferry], 2015).
Adams, Philip, *Daring to Defy. Port Talbot's War Resistance: 1914-18* (Philip Adams, 2016).
Anon, *Pit Pony* (booklet, National Coal Board, n.d.).
Anon, *The Rotherham Annual 1915-1917* (Rotherham Archives & Local Studies, n.d.).
Arnold, Catharine, *Pandemic 1918* (Michael O'Mara Books, 2018).
Arnot, Robin Page, *The Miners: Years of Struggle. A History of the Miners' Federation of Great Britain (from 1910 onwards)* (Allen & Unwin, 1953).
Arnot, Robin Page, *A History of the Scottish Miners From the Earliest Times* (Allen & Unwin, 1975).
Arnot, Robin Page, *South Wales Miners. A History of the South Wales Miners' Federation 1914-1926* (Allen & Unwin, 1975).
Ashcroft, Michael, *Victoria Cross Heroes* (Headline, 2006).
Atterbury, Paul, *Antique Roadshow. World War One in 100 Family Treasures* (BBC Books, 2014).
Barlow, Robin, *Wales and World War One* (Gomer, 2014).
Barnes, David, *Black Mountains* [The Recollections of a South Wales Miner] (Dinas/Y Lolfa, 2002).
Barrie, Alexander, *War Underground. The Tunnellers of the Great War* (Tom Donovan, 1961).
Barton, Peter, Peter Doyle and Jonathan Vandewalde, *Beneath Flanders Fields. The Tunnellers' War 1914-18* (Spellmount, 2004; The History Press, 2010).
Baxter, Shirley, and Hilary McAra, *A Life Too Soon Gone. Sheffield General Cemetery and the Great War* (Sheffield General Cemetery Trust, 2016).
Baylies, Carolyn, *The History of the Yorkshire Miners 1881-1918* (Routledge, 1993).
Beachill, James, *The Terrible Pit Disaster* (Conisbrough Local History Society, n.d. c.2011–12).
Bell, David, *Derbyshire Heroes* (Countryside Books, 2004).
Bellamy, Joyce, and John Saville (eds), *Dictionary of Labour Biography*, vols 1–6 (Macmillan, 1972–9).
Benson, John, *British Coalminers in the Nineteenth Century* (Holmes & Meir, 1980).
Benson, Richard, *The Valley* (Bloomsbury, 2015).
Bilton, David, *Allied POWs in German Hands 1914-18* (Pen & Sword, 2019).
Blunden, Edmund, *The Poems of Wilfred Owen* (Chatto & Windus, 1931).
Blunden, Edmund, *The Undertones of War* (R. Cobden-Sanderson, 1928).
Brazier, Kevin, *The Complete Victoria Cross* (Pen & Sword, 2010).
Bridgland, Tony, and Anne Morgan, *Tunnel-Master & Arsonist of the Great War. The Norton Griffiths Story* (Leo Cooper, 2003).
Brookes, Gill, *Kingstone Remembers the Great War 1914-1918* (Kingstone Heritage Group, 2014).
Bulman, H.F., and Frederick P. Mills, *Mine Rescue Work and Organization* (Crosby Lockwood, 1921).
Busby, Mick, *They Shall Not Grow Old. Rawmarsh and Parkgate Men Who Fell in the Great War During 1914 and 1915* (n.d., c.2016, self-published, Rotherham).
Campbell, Alan, Nina Fishman and David Howell (eds), *Miners, Unions and Politics 1910-1947* (Routledge, 1996).
Church, Roy, *The History of the British Coal Industry* [Volume 3, 1830-1913: Victorian Pre-eminence] (Oxford Clarendon Press, 1986).
Compton, Ann, *The Sculpture of Charles Sargeant Jagger* (Lund Humphries, 2004).
Cooksey, Jon, *Barnsley Pals* [The 13th and 14th Battalions York & Lancaster Regiment] (Pen & Sword, 1986/2008).
Cowen, Torqui, *Labour of Love. The Story of Robert Smillie* (Neil Wilson Publishing, 2011).
Dearne Local History Group, *'More Dearne Days Remembered'* (booklet, 1988).

Derycke, Karen, and Lee Inglebrecht, *Memorial Museum Passchendaele 1917 Visitor Guide* (2015).
Downes, Eddie, *Yorkshire Collieries 1947-1994* (Think Pit, 2016).
Drinkall, Margaret, *Rotherham in the Great War* (Pen & Sword, 2014).
Duncan, Robert, *The Mineworkers* (Birlinn, 2005).
Elliott, Brian, *Barnsley 1890s-1990s* (Sutton Publishing, 1999).
Elliott, Brian, *Images of the Past. Coal Mine Disasters in the Modern Era c.1900-1980* (Pen & Sword, 2017).
Elliott, Brian, *South Yorkshire Mining Disasters*, 2 (Wharncliffe/Pen & Sword, 2009).
Elliott, Brian, *Tracing Your Coalmining Ancestors* [A guide for family historians] (Pen & Sword, 2014).
Elliott, Brian, *Yorkshire Mining Veterans in Their Own Words* (Pen & Sword, 2005).
Evans, Geoffrey, *The Men Who Marched Away. Unlocking the Cynon Valley War Memorials* (Cynon Valley History Society, 2014).
Faulks, Sebastian, *Birdsong* (Hutchinson, 1993).
Fisher, Elaine H., *Requiem for Will* (E.H. Fisher/Harrison Military, 1997).
Fordham, Dave, *Maltby Main Colliery* (Fedj-el-Adoub, 2015).
Francis, Hywell, and David Smith, *The Fed. A History of the South Wales Miners in the Twentieth Century* (Lawrence & Wishart, 1980).
Freemantle, Michael, *Gas! Gas! Quick Boys! How Chemistry Changed the First World War* (History Press, 2014).
Gliddon, Gerald, *VCs of the First World War: Arras and Messines* (History Press, 1998/2012).
Gliddon, Gerald, *VCs of the First World War. Cambrai 1917* (History Press, 2012).
Gliddon, Gerald, *VCs of the First World War. Somme 1916* (History Press, 2012).
Goodman, Martin, *Suffer & Survive. Gas Attacks, Miners' Canaries, Spacesuits and the Bends. The Extreme Life of Dr J.S. Haldane* (Simon & Schuster, 2007).
Graves Robert, *Goodbye to All That* (Robert Cape, 1929; Penguin, 1960).
Grieve, Captain W. Grant, and Bernard Newman, *Tunnellers. The Story of the Tunnelling Companies, Royal Engineers, During the World War* (Herbert Jenkins, 1936/Naval & Military Press, 2006).
Griffin, Alan R., *The Miners of Nottinghamshire 1914-1944* (George Allen & Unwin, 1962).
Griffiths, Jim, *Pages from Memory* (Dent, 1969).
Groom, Brian, *Northerners. A History* (Harper North, 2022).
Hamilton, Robert, *Victoria Cross Heroes of World War One* (Atlantic Publishing, 2017).
Harrison, Michael, *High Wood* (Pen & Sword, 2017).
Hemingway, Ernest, *For Whom the Bell Tolls* (Jonathan Cape, 1940).
Henderson, D.V., *Heroic Endeavour* [A Complete Register of the Albert, Edward and Empire Gallantry Medals and How They Were Won] (J.B. Hayward, 1988).
Hetherington, Andrea, *British Widows* (Pen & Sword, 2018).
Hibberd, Dominic, *Wilfred Owen: A New Biography* (Orion, 2003).
Hill, Alan, *The South Yorkshire Coalfield. A History and Development* (Tempus/The History Press, 2001).
Hodgkiss, Winifred Haward, *Two Lives* (Yorkshire Arts Circus, 1983).
Horner, Arthur, *Incorrigible Rebel* (Macgibbon & Key, 1960).
Humphries, Steve, and Richard van Emden, *All Quiet on the Home Front* (Headline, 2003; Pen & Sword, 1917).
Jackson, Robert, *The Prisoners 1914-18* (Routledge, 1989).
Jenkins, R.T., E.D. Jones and Brinley F. Roberts (eds), *Dictionary of Welsh Biography 1941-70* (Honourable Society of Cymmrodorion, 2001).
Jevons, Herbert Stanley, *The British Coal Trade* (Kegan Paul Trench Trubner, 1915 [and David & Charles Reprints, 1969]).
Johnson, Malcolm (ed), *Miners' Battalion. A History of the 12th* [Pioneers] *King's Own Yorkshire Light Infantry 1914-1918* (Pen & Sword, 2014).
Jones, Heather, *Violence Against Prisoners of War in First World War Britain, France and Germany* (Cambridge University Press, 2011).
Jones, Simon, *Underground Warfare 1914-1918* (Pen & Sword, 2010).
Kramer, Anne, *Conscientious Objectors of the First World War. A Determined Resistance* (Pen & Sword, 2014).

Lawrence, Bill, *From Pit Town to Battlefields 1914-1916* (LEB Books, 2015).
Lewis-Stempel, John, *The War Behind the Wire* (Weidenfeld & Nicolson, 2014).
Lloyd, John Edward, and R.T. Jenkins, *Dictionary of Welsh Biography Down to 1940* (William Lewis, 1959).
Lomax, Scott, *The Home Front* [Derbyshire in the First World War] (Pen & Sword, 2016).
Lynch, Tim, *Wakefield in the Great War* (Pen & Sword, 2017).
MacFarlane, Jim (J.E.), *The Bag Muck Strike. Denaby Main, 1902-1903* (Doncaster Library Service, 1987).
Mace, Martin, and John Grehan, *Slaughter on the Somme. 1st July 1916. The Complete War Diaries of the British Army's Worst Day* (Pen & Sword, 2016).
Manise, Mick, *Stories From the Great War* [Volume 1]. The Pooley Miners (Grosvenor House, 2018).
Mansergh, Ruth, *Whitehaven in the Great War* (Pen & Sword, 2017).
Marsh, Arthur and Victoria Ryan, *Historical Dictionary of Trade Unions* (Volume 2) (Gower, 1984).
McHenry, Iain, *Subterranean Sappers. A History of 177 Tunnelling Company RE From 1915 to 1919* (Uniform Press, 2015).
McLean, Iain, *Keir Hardie* (Allen Lane/Penguin, 1975).
McMillan, Sheila, *Corporal George Bradley and the Star and Garter Home for Disabled Soldiers and Airmen* (Little Book Club, 2023).
McMillman, Bruce, *Managing Domestic Dissent in First World War Britain* (Frank Cass, 2000).
Mellor, Michael (et al., ed.), *A Nation's Tribute* (CHAP : Chase Arts for Public Spaces, 2021).
Mills, Catherine, *Regulating Health and Safety in the British Mining Industries, 1800-1915* (Ashgate, 2010).
Moffat, Abe, *My Life With the Miners* (Lawrence & Wishart, 1965).
Moore, Robert, *Pitmen, Preachers and Politics* (Cambridge University Press, 1974).
Morgan, Kenneth O., *Keir Hardie, Radical and Socialist* (Orion, 1997).
Mullins, Claud, *The Leipzig Trials, an Account of the War, Criminal Trials and a Study of German Mentality* (Geco Ecco, 2012 [first published 1921]).
Murray, Joseph, *Gallipoli As I Saw It* (William Kimber, 1965).
Murray, Joseph, *Call to Arms* (William Kimber, 1980).
Norman, Terry, *The Hell They Called High Wood: The Somme 1916* (Pen & Sword, 2009).
Oldfield, Paul, *Victoria Crosses on the Western Front. 1917 to Third Ypres 27 January 1917-July 1917* (Pen & Sword, 2016).
Pearce, Cyril, *Comrades in Conscience. The Story of an English Community's Opposition to the Great War* (Frank Boutle, 2014).
Pearce, Cyril, *Communities of Resistance. Conscience and Dissent in Britain during the First World War* (Frank Boutle, 2020).
Powell, Adam, *Soldiering On. British Tommies After The First World War* (The History Press, 2019).
Ramsden, John (ed.), *The Oxford Companion to Twentieth-Century Politics* (Oxford University Press, 2002).
Redmayne, Sir Richard A. S., *The British Coal-Mining Industry During the War* (Oxford Clarendon Press, 1923).
Richards, Frank, *Old Soldiers Never Die* (Faber, 1933; Parthian, 2016).
Richardson, Matthew, *Deeds of Heroes. The Story of the Distinguished Conduct Medal 1854-1993* (Pen & Sword, 2012).
Richter, Donald, *Chemical Soldiers: British Gas Warfare in World War One* (Pen & Sword, 2014).
Robinson, Philip, and Nigel Cave, *The Underground War. Vimy Ridge to Arras* (Pen & Sword, 2011).
Rodgers, Alice et al, *God's Coal. The Church's mission amid development and change* (Church of England [booklet], 1990).
Rosen, Greg (ed), *Dictionary of Labour Biography* (Politico, 2001).
Senior, Michael, *The Soldiers' Peace. Demobilizing the British Army 1919* (Pen and Sword, 2018).
Shannon, Stephen, *Beyond Praise: The Durham Light Infantrymen Who Were Awarded the Victoria Cross* (County Durham Books, 1998).
Smith, Michael, and Kay Valentine, *Darfield Remembers The First World War* (Darfield History Society, 2016).

Snelling, Stephen, *VCs of the First World War: Passchendaele* (History Press, 1998/2012).
Steward, Graham, and John Sheen, *Tyneside Scottish: 20th, 21st, 22nd, and 23rd (Service) Battalions of the Northumberland Fusiliers* (Pen and Sword, 1999).
Sheen, John, *Tyneside Irish: 24th,25th,26th and 27th (Service) Battalions of the Northumberland Fusiliers* (Pen and Sword, 2010).
Streets, J.W., *Undying Splendour* (Erskine McDonald, 1917).
Stuart Wortley, Rothesay, *Letters of a Flying Officer* (Sutton, 1982).
Supple, Barry, *The History of the British Coal Industry. Volume 4, 1913-1946* [The Political Economy of Decline] (Oxford Clarendon Press, 1987).
Tomaselli, Phil, *Givenchy in the Great War* (Pen & Sword, 2016).
Thompson, Ceri, *Harnessed* [Colliery Horses in Wales] (National Museum Wales, 2008).
Turner, P.W., and R.H. Haigh, *Not For Glory* [re; Gilbert Hall] (Robert Maxwell, 1969).
Vack, Dale le (ed), *Stretcher Bearer* (Lion Hudson, 2013).
Van Emden, Richard, *Boy Soldiers of the Great War* (Headline, 2005; Pen & Sword, 2021).
Van Emden, Richard, *Britain's Last Tommies* (Pen & Sword, 2005).
Van Emden, Richard, *Missing: the Need for Closure after the Great War* (Pen & Sword, 2020).
Van Emden, Richard, *Prisoners of the Kaiser* (Pen & Sword, 2009).
Van Emden, Richard, *The Somme. The Epic Battle in the Soldiers' Own Words and Photographs* (Pen & Sword, 2016).
Van Emden, Richard, *The Road to Passchendaele* (Pen & Sword, 2017).
Westwood, Frank, and Andrew Featherstone, *Rotherham and District War Memorials* (privately published, 2003 [available in Rotherham Archives and Local Studies Library]).
Whitworth, Alan, *Yorkshire VCs* (Pen & Sword, 2012).
Wilkinson, Oliver, *British Prisoners of War in First World War Germany* (Cambridge University Press, 2017).
Williams, J.E., *The Derbyshire Miners* (George Allen & Unwin, 1962).
Williams, W. Alistair, *Heart of the Dragon. The VCs of Wales and the Welsh Regiments, 1914-82* (Bridge Books, 2008).
Wood, Ritchie, *Miners at War 1914-1919. South Wales Miners in the Tunnelling Companies on the Western Front* (Helion & Company, 2017).
Wright, Christopher, and Glenda Anderson (eds), *The Victoria Cross and the George Cross. The Complete History* [Volume 2, First World War], (Methuen & Victoria Cross Association, 2013).
Wright, Ian, *Coal On One Hand, Men on the Other. The Forest of Dean Miners' Association and the First World War 1910-1920* (Bristol Radical History Group, 2014).
Wylly, Colonel Harold Carmichael, *York and Lancaster Regiment 1758-1919* (Frome & London, 1930, [via Rotherham Archives and Local Studies]).

INDEX

Abberville Communal Cemetery 225
Aber Valley 74
Abercam 16
Abertillery 136, 148
Abertridwr 74
Ablett, Noel 208
Abram 54
Absolutist 151
Accrington Pals 108
Ackroyd, Walter 34
Acton, Abraham VC 40, 43
Adams, Philip 144
Addison, Dr 231
Aisne 39, 40
Aldershot 183, 231
Aleppo 203
Alexandria 27
Allied PoWs in German Hands 1914-1918 (book) 198
All Quiet on the Western Front (film) 155
Allen, Rev Montague 157
Allenby, Sir Edmund 67
Alnwick 175
Ambulance Brigade 93
ambulances (field) 31, 94, 150, 165
ambulances (war) 50-2
Amiens (battle) 214
Ammington 65
Angus, William VC 55-6
Ansty, WJ 148
Antcliffe, Ernest 186-8
anti-German feeling 24, 50
Antiques Roadshow (BBCTV) 150, 223
Antwerp 24, 87
Arksey Cemetery 175-6
Arleux 97
Armentieres 27, 42, 174
Armistice 35, 76, 98, 99, 102, 195, 200, 208, 213, 230, 1 231, 239, 240, 241, 242
Armley (prison) 145
Arras 158, 160, 181, 182, 188, 212
Arrowsmith, Wilfred 149
Ashfield 168
Ashington 1, 22, 45, 181, 253
Ashton, Frank 161
Atterbury, Paul 150, 187, 223
Auton Stile 146
Aveluy Wood 94

Badman, Fred 149
Bailleul Communal Cemetery Extension 46
Bakewell, John 50

Ball, Henry 91-2
Ball, John 96
Banister, Wilfred 90
Baptist 148, 149
Barham HMS 80
Barlinnie (prison) 149
Barnard Castle 67
Barnby Dun 157
Barnes, David 74
Barnsley 24, 25, 32, 41, 43, 49, 66, 79, 100, 138, 141, 142, 143, 150, 158-62, 173, 179, 180, 193, 225, 236, 237, 241, 242-5, 246, 250, 251
Barnsley Chronicle 32, 40, 57, 158, 159, 161, 162, 196, 197
Barnsley Pals (also see 13/14 York and Lancasters) 108, 109, 138, 142, 158, 160, 162, 246, 248, 249, 250, 251
Barrs, Thomas 184-5
Barter, Frederick VC 56
Barton, Peter 15, 67, 71, 73
Bayley, Henry Dennis 50-1
Beaumont Hamel 178
Bedlington 22
Bedson, William Henry 122-3
Beacon Museum, The 40
Beesley, William VC 214
Beevers, Lavinia 39
Belgium refugees 24-5, 41, 49
Bell, George 194-5
Bell, Joseph 159
Bell, Thomas Frederick 159
Bellewaerde Ridge 68
Bellicourt 239
Beneath Flanders Fields (book) 67
Benn, Edward VC 213-4
Benson, Richard 236
Benson, Walter 236
Berger, Edward 155
Bethune 61, 64, 65
Bilton, David 191, 198
Binney, William 96
Birchall, Charles 88
Birmingham 172
Blackpool 163
Black Mountains (book) 74
Blaina 28
Blair, Robert Richmond Curwin DSO 43
Bleanavon Iron & Coal Company 120
Blenkinsop, Mr J 84, 87, 88
Bishop Auckland 136
Blunden, Edmund 63, 65, 205
Bodmin 136

Boer War 19, 33, 59, 115
Bolton 25, 234
Bolton-upon-Dearne 222, 255
Boulogne 47
Bowes, John and Partners 23
'boy soldiers' (see miners underage)
Brace 'scheme' (conscientious objectors) 20, 48, 140, 148-9, 150
Bradford 137
Bradford, George 136
Bradford, Roland Boys VC 135-6
Bradley, George 220-21
Bradley, John 67
Brancepath 228
Briddon, Frederick 87
Bridgend 183
Bridges, Percival 223-4, 235
Bridgland, Tony 67
Briggs, James 91
British Coal-Mining Industry During the War (book) 189
British Expedition Force (BEF) 34, 40, 77, 99, 120, 185
British War Medal 31, 35, 38, 77, 78, 90, 142, 181-2, 223, 229
British Widows (book) 158
Brodie, William 26-7
Brocklesby, Robert C 102
Broke HMS 86
Brooks, Oliver VC 56
Brown, James 181-2
Brown, Philip 47
Bryan, Thomas VC 175-6
Buckingham Palace 40, 47, 50, 56, 95, 180, 218
Bucquoy 214, 216
Bullen, James 95
Burnley 147
Burnopfield 29
Butler, William Boynton VC 180
Bye, Robert James VC 176-7

Caberet-Rouge Cemetery 160
Cable, John Richard 145-6
Caerphilly 74, 148, 238
Calliope, HMS 29
Calthrop, Guy (also see Coal Controller) 189, 208, 231, 233, 235
Cambrai 155-6, 179, 228, 247
Cambrian Combine 44, 199
Cambrin Military Cemetery 65
Campbell, George 67
canary (birds) 170, 172
Cann, Thomas 145
Card, Oliver 192
Cardiff 21, 41, 139, 183-4, 223
Cardiff (prison)149
Carlton (Barnsley) 185, 196
Carluke 55-6
Carnarvon (prison) 149

Carrie, William 147
Cassels, Geoffrey Rowley 66
Castleford 176
Caterpillar Crater 166
Cecil, Evelyn 227
certificates of exemption 139
Chafer, George William VC 95, 137
Chapel-en-le-Frith 93
Charnock, Walter 92
Chase Arts for Public Spaces (CHAPS) 257-8
Chatham 65, 188
Chatt, George Auty 67-9
Chavonne (France) 39
Chemnitz (prison camp) 195
Chesterfield 45, 151, 186, 225, 234
Chili Trench Cemetery (Gavrelle) 97
Choques Military Cemetery 229
Christadelphians 147, 148
Christmas Day Truce 1914 29
Church Hesley 215
Churchill, Winston MP 248, 257
Cite Bonjean Military Cemetery 27
Clarke, Edwin 99
Clarke, Francis V 239
Clauson, George 258
'clay-kickers' 57
Clifford, Arthur Bernard 172-5, 205
Clifford, Hubert Aubrey 198-9
Clifford, Walter 173
Clifton and Kersley Coal Co 23
Coal-Controller 17, 189, 208, 231, 232, 233
Coal Mining Organisation Committee (CMOC) 43, 44
coalfields:
    Bristol 207
    Cannock Chase 48, 142
    Cumberland 48
    Durham 45
    East Midlands 13
    Forest of Dean 132, 174, 207
    Kent 13, 24, 48
    Lancashire & Cheshire 13
    Midland 48
    North East 13
    North Wales 13
    Scotland 13
    Southern 48
    South Wales 13, 20, 48, 144
    Staffordshire 71
    Yorkshire, Notts & Derbys 13, 45, 48
coal-cutting machines 13
coal owners 13, 42, 44
collieries:
    Allerton Bywater 218-19
    Altham (Moorfield) 149
    Ashington 253
    Ashton Moss 137

Index • **279**

Askern 157, 176
August Victoria 193
Backworth 39
Bank Hall 228
Barnsley Main 49, 196, 249, 250
Barrie, Alexandre 67
Barrow 158
Bedwas 202
Beeston 39
Bentinck 52
Bentley 176, 186
Birchwood 204
Black Vein 209
Blackhall 22
Blackie 27
Bonds Main 177
Brancepath 121
Brandon 158
Braysdown 224
Brodsworth Main 24, 111
Browny 115
Bryn Hall 54
Cadeby 121, 123, 200, 221
Chislet 2
Cortonwood 78, 161, 254
Cwmtillery 136-7
Dalton Main (see Silverwood)
Deep Dyffryn 176
Denaby Main 24, 31, 123, 194, 195, 200, 221
Digby 241
Dinnington 186
Dodworth 249
Elliot 148, 167, 200
Elsecar Main 33
Emma 110, 112, 115, 117
Exhall 52
Exhill 77
Farrington Gurney 149
Ferndale 244
Firbeck 177
Frederich der Grosse 202
Grassmoor 49
Greenside 1, 110, 112
Grimethorpe 6, 34-5, 51, 159, 197
Hampstead 173
Hapton Valley 227
Harrington (Lowca) 40
Harworth 24
Haunchwood 215
Hebburn 145
Hem Heath 52
Hemsworth 200
Hickleton Main 37-8, 222, 224, 252-3
Houghton Main 110, 138, 160, 161, 193, 220
Kiveton 150
Lady Lewis 209

Littleburn 48
Llanbradach 232
Maltby Main 60, 186, 188
Mansfield 53, 234
Manvers Main 128, 145, 253
Markham (Derbys) 111, 177, 178
Markham Main (Doncaster) 24
Marley Hill 65
Minnie (see Podmore)
Monckton Main (New) 51, 162, 179-80, 197, 246-8
Monk Bretton 159, 249
Mossfield 143
Mosside 241
Moston 80
New Broughton 71
New Hucknall 182
Niddrie 39
Norton (Yorks)176
North Gawber 251
North Seaton 45
Norton Hill 79
Oakdale 120
Podmore Hall 15, 52, 174, 204-06
Pooley Hall 65
Primrose Hill 253
Rainford 36
Rob Royd 249
Rossington Main 157
Rotherham Main 19, 42, 192
Roundwood (see Silverwood)
Ryhope 45
Shipley 216
Silksworth 146
Silverhill 168
Silverwood 82-105, 137, 145
South Garesfield 22, 29
St Hilda's 40
St John's 23, 60
Stargate 110
Strafford 249, 250
Sutton 168
Thorne 24
Thornley 46
Thrislington 23
Thrybergh Hall 252
Thurcroft Main 183
Universal 73
Warsop 177
Wath Main 30-31, 251
Welbeck 177
Wellington Pit 17
Wharncliffe Woodmoor 57, 81, 185, 249, 250
Wheatley Hill 47
Wheldale 176
Whitburn 145
Whitwood 176

Willington 121
Windsor 74, 76
Wombwell 250
Yorkshire (Edlington) Main 253
Colincamps 28
Colliery Guardian 20, 48, 146, 164
Colliery Recruitment Courts (CRCs) 139, 140, 150, 151, 164, 207
Collingridge, Frank 250
Collingwood, Taylor 195-6
Collins, John VC 180
Communist Party of Great Britain 242
Conisbrough 24, 123, 221
conscientious objectors (also see Brace scheme) 9, 139, 144-152, 240-41
Continental Times 196
Conway, Thomas 138
Cooksey, Jon 58
Coventry 215
Cowen, J Hunter 66
Cox, Joe 99
Craiglockhart (hospital) 205
Crawcrook 115
Croix de Guerre (medal) 51, 180
Crooks, James Ellis 221-22
Cropper, Cecil 120
Cross, CH 148
Cross of the Order of St George 53
Crystal Palace 87, 136, 188
Cudworth 225
Cuinchy 61-3, 65, 66
Cunnane, Philip 88
Cwmafan 144

Daedalus HMS 188
Dalgas, Agner 240
Dalton 88, 91, 93, 95, 101
Dalton, Enoch 60-5
Dalton Main Company 88, 91, 95, 101
Dardanelles 41
Darfield 160, 194, 220, 238
Dartmoor 140, 144-8, 150, 151
Darton 160, 251
Davies, Eli 64
Davies, Henry 144
Davies, James Llewellyn VC 180
Davies, John Henry 57-8
Davies, Joseph John VC 137
Davies, Thomas Edward 167
Davis, Frank 99
Davis, Sir Robert 173
Deakin, Arthur 150
Dearne Valley (war memorial) 255
DeComyn, Andy 257
Defence of the Realm Act (DoRA) 145, 156
Delville Wood 107, 137

demobilization 104, 231-45
Dent, Harry 99
Derby Scheme 43, 45
Derbyshire Miners' Association 50
deserter 80, 150
Dibbo, Edward 93-4
Dick, William 147
Dinsdale, Ernest 150
Distinguished Conduct Medal (DCM) 8, 29, 60, 62, 88, 102, 150, 197, 199, 216, 217, 221, 222, 248, 249, 250
Distinguished Flying Medal (DFM)186
Distinguished Service Medal (DSM) 87
Distinguished Service Order 120, 121
Doberitz camp (prison) 194-5
Dobson, Frederick William VC 39
Dodworth 250
Doncaster 24, 45, 49, 121, 157, 176, 194, 257
Dowd, George 99
Dowlais 232, 241
Dover 192, 203
Drabble, Sydney 89-90
Dranoutre Military Cemetery 43
Drury, Len James 100
Dudley 142
dugouts 210-12
Dunkirk 188
Dunning, Richard 117
Dunsire, Robert VC 56
Durham Big Meeting 46
Durham Chronicle 46, 146
Durham miners 42, 43, 46, 47, 65, 67
Durham Miners' Association (DMA) 22, 52, 145, 148, 149
Durham prison 148
Duxbury, Samuel 96
Dyson, Walter 19

Eades, John 50
Ebbw Vale Company 218
Edinburgh 39
Edward Medal 17, 173
Edwards, Arthur 120
Edwards, 'Nes' 148
Edwards, Philemon 149
Edwards, Wilfred VC 180
Egerton, Albert VC 180
Egypt 34, 39
Eight Hours Act (1908) 44
Elliott, Fred 28
Ellis, John 96
Ellison, George Edwin 230
Else, Stanley 94
Elsey, Herbert 96
Elswick 29
Emden, Richard Van 36, 158, 191
Epsom Hospital 141, 236
Epworth 137

Index • **281**

Esh Winning 68, 154
Etaples 226, 227, 229
Evans, G 148
Evans, William Rees 199
Everson, Rev Hubert Rouse 84, 92
Exeter prison 149
Exley, Charles 141

Faulkes, Sebastian 6, 15
Featherstone, Alfred Donavon 97, 98
Featherstone, Andrew 98
Featherstone, Bernard 97-8
Featherstone, Cyril 97
Featherstone, George 97
Festubert 40
Finn, James Henry VC 136
Finny, John 99
Firth, Sidney 167
Fisher, Edwin 93
Fisher, Herbert 87
Fisher, William 73-4
Fitch, Harold 31
Flers Courcelette 75, 107
Fletcher, AFG (chaplain) 46
Fleuss, Henry 173
forced labour 193-5
Fort, Cyril 249
Frankfurt 198
Freeth, HE (artist) 227
Frickleton, Samuel VC 180
Fuller, Wilfred Dolby VC 53

Gaitskill, Hugh MP 248
Gallipoli 30, 41, 54, 79, 93, 99, 110, 136, 178, 182, 256
Gardiner, James 117-20
Gardner, William 200
gas 31, 39, 59, 75, 79, 96, 169, 183-4, 210, 220, 238, 256
Gaskell, Randell 90
Gateshead 65
Gee, Ernest 92
George V, King 83, 90, 136, 176, 218
Giessen camp (prison) 198, 199
Gill, Len 249
Givenchy 16, 46, 55, 56, 61, 63, 65, 91
Glasgow Celtic FC 55-6
Glover, Harry 159-60
Glynn, John 158
Goldthorpe 50, 255
Gordon, Alfred 74-5
Gottingen camp (prison) 195
Greasborough 100
Great Houghton 160
Greaves, Fred VC 177-8
Green, Corporal CE 193
Green, Horace 92-3
Greer, William 61

Gregg, William VC 214
Grimshaw, John Elisha VC 5, 54-5
Goodbye to All That (book) 61
Good Hope, HMS 30
Graves, Robert 29-30, 61-3, 65
Gray, Edward 95
Greaves, Ernest 96
Grimethorpe 34,
Grimsby 32
Gustrow (PoW camp) 89

Hackett, William VC 15, 63
Haig, Sir Douglas 211-2, 235
Haldane, Dr John Scott 171-2
Hall, Charles Sydney 181
Hall, Gilbert 33-4
Hall, Harry (Barnsley) 109-10
Hall, Harry (Dalton) 95
Haltwhistle 228
Hamburg, 194
Hardie, Keir 26
Harper, James William 242
Hartley 22
Hastie, John 147
Haston, Henry 151
Hatt, Edward Beach 75
Hawke, HMS 109
Hawthorne Ridge 124, 200
Howarth, John 162
Haynes, John 91
Heanor 215
Heaviside, Michael Wilson VC 47, 180
Hebuterne 214
Hemingway, Ernest 153
Hemmingfield 78
Hemsworth 164, 247, 248
Hepburn, Allan 187
Hepburn, William Clay 120
Heppinstall, Fred 96
Hetherington, Andrea 158
Hetton 46
Hewitt, Sir Joseph 23, 32, 246, 248
Hexham 163
Heynen, Karl 202-3
Hickleton, Viscount 23
Hickman, Ben 99
High Wood 107
Hill, Albert 137
Hill, Charles Chambers 221
Hill 60 41, 56, 57, 57, 58, 66, 67, 70, 89, 122, 159, 166, 198, 210
Hill 63 210
Hill 70 89
Hindenburg Line 104, 155, 180
Hodge, Frank 233
Hodges, W (Gunner) 96

Hodgkiss, Louis 80
Hodgkiss, Winifred Howard 80
Hogg, Catherine 117
Hogg, Richard 115, 117
Hogg, William Armstrong 117
Holbrook, William 99
Holmes, Albert Henry 241-2
Holmes, George 96
Holmes, Horace Edwin 246-8
Hooge 66, 67, 184
Hooton Pagnell 237
Hop Store Cemetery (Ypres) 38
Horner, Arthur 148
Horner, Charles 226
Hoskins, GR 66
Howe, Richard 142
Hoyland 33, 138
Hudspeth, Henry Moore 120-22, 123
Huit, Major R 192
Hull, Ernest 154
Hullock 199
Hullock, Ernest 38-9
Hunslet 38
Hunt, Arthur 89

Imperial War Museum 138, 176, 190, 256
Ince 182
Independent Labour Party (ILP) 144, 149
Inflexible, HMS 81
influenza pandemic 9, 199, 229, 233, 241, 242
Irving, Henry J 1
It's a long way to Tipperary (popular song) 32

Jackson, Robert 190
Jagger, Charles Sergeant MC 256-7
Jagger, Enoch 256
Jarrow 145, 245
Jenkins, Dai 183, 222-3
John Brown Company 83, 102
John Copse 27, 109
Johnson, Rev Ernest Edward 157
Johnson, Hugh 52
Jones, Ernest 104-5
Jones, Francis 104
Jones, Heather 191
Jones, Leslie 104-5
Jones, Simon 71
Jones, William James 81
Jowett, Herbert 95
Jutland (battle) 80-1, 86

Keep the Home Fires Burning (popular song) 44
Keeton, CE 196
Kelly, Bartholomew 148
Kelly, Frederick 96
Kelly, James 148

Kenealy, William Stephen VC 54-5
Kenny, Thomas VC 47, 53, 55
Kenworthy, John Dalzell 40
Killamarsh 177
Kirk Bramwith 157
Kirk, Horace 224-5
Kirk, Samuel 224-5
Kirkcaldy 22
Kirkintilloch 147
Kitchener, Herbert 1st Earl 11, 18, 19, 22, 48, 184, 230
Knowles, Andrew and Sons 22
Kut (seige) 203

La Boisselle 71-3, 113-5, 116, 119
Lager Dulmen camp (prison) 193
Lambley, Joseph Thomas 100
Lancaster Convalescence Home (Barnsley) 237-8
Lancaster, Edward 237
Launchbury (family) 76-8
le Cateau 157, 193, 239
Leatham, Claud 160-61
Leeds 50, 121, 145, 150, 167, 218, 219, 253
Leet, Alfred 19
Legion of Honour 52
Leipzig War Trials 202-3
Lens 64
Lewis, David 149
Lewis, J Dyer (mines inspector) 139, 152, 206
Lewis-Stemple, John 191
Lightfoot, David 193
Light Lines (art installation) 143
Lindley, Frank 109
listening duty 41, 64, 68
Lister, John 147
Lintz 29
Liverpool (prison) 145
Llanelly 149
Llewellyn, Len 44
Lloyd George, David MP/PM 13, 44, 153, 156, 207, 233, 242-4, 250
Lloyd, William Arthur 71
Lochnavar Crater (memorial site) 71, 111, 113, 116-7, 119, 228
London 49, 60, 85, 136, 151, 173, 188, 197, 199, 231, 257
London Gazette 199, 210, 214
Lone Tree Crater (see Spanbroekmolen)
Loos 31, 36, 39, 41, 68, 78, 177
Loos-en-Gohelle (St Patrick's Cemetery) 64
Lothians 48
Lothian Miners' Association 22
Louvain 24
Loversall Hall 237
Ludworth 163
Lusitania RMS 44
Lys (battle) 211, 222

Index • **283**

Main, Richard 145-6
Mainwaring, Tal 145
Maltby 60, 63, 64, 186, 188
Mametz Wood 108, 169, 217
Manchester 25, 80, 234
Mannise, Mick 65
Mansfield 23, 111, 234, 239
Mariner, William VC 56
Markham, Sir Arthur 53, 111
Marne 40
Marriott, William 102
Marshall, John 99
Martin, James 56
Maryport 48, 213
Masbrough 99, 102
Mason, Roy MP 248
Matthew Copse 28
McAulay, John VC 180
McHenry, Iain 67, 240
McIver, Hugh VC 218
McKenna, Reginald 43
McMillan, Sheila 220-21
McNally, William VC 213
Meaulte 137
Mecklenburg camp (prison) 194
Medaille Militaire 180, 218, 247
Meek, James 238
Mellor, Michael 258
Memorial Plaque 90, 98, 117
Mendes, Sam 155
Menin Gate Memorial 58, 70, 154, 159, 167
Merthyr 208, 241
Messines 66, 122-3, 153-4, 164-5, 166, 167, 210, 212, 219, 260, 210, 212
Mexborough 30, 156, 195, 254
Middlemiss, George Thomas 65-6
Middlewood Hall 238
Miles, Francis VC 218
Military Cross 51, 52, 67, 100, 120, 122, 205, 256
Military Medal (MM) 29, 87, 94, 95, 96-7, 100, 167, 178, 210, 213, 216, 221, 222, 251
Military Service Act 124, 139, 145, 146, 164, 206
Military Service Tribunal (MST) 140
Milnes, Enoch 96
miners:
    accidents 8, 52, 59, 95
    absenteeism 44, 49, 164, 189, 208, 220, 234
    'butty' system 19, 244
    'comb-out of' 150, 151, 163-4, 189, 206-7
    disabled 235
    disasters 17, 123
    Senghenydd 73-4
    Wellington 17, 43
    Minnie/Podmore Hall 204-6
    disputes/lockouts/strikes 94, 233
    fatalities 14, 15, 90-1
    haulage hands 20
    pit ponies 19
    pony drivers 14, 90, 96, 102, 141, 192
    Price Lists 243
    underage 35-6, 99, 108-9
Miners at War 1914-1919 (book) 69, 166
Miners Federation of Great Britain (MFGB) 11, 48, 189, 207-8, 232, 242
mines inspectorate 51, 122
mines rescue 44, 67, 169-75, 195, 204-6
mines rescue stations 16-17
    Brierley 51
    Cowdenbeath 16
    Crumlin 173
    Doncaster 16
    Dudley 172
    Elswick 16
    Howe Bridge 16
    Mansfield 16
    Tankersley 16, 173
Mining engineers 120-22
Mining Institute (NEIMME) 50, 121, 172
Minto, George 148
Miree, Samuel 101-2
Missing (book) 158
'Mission churches' 84
Molyneux, John VC 180
Monmouth, HMS 30
Mons 20, 99
Moore, Robert 95
Morgan, Anne 67
Morgan, David Watts 44-5
Morris, Joseph 89
Morrison, James 46
Mottram, Thomas 26, 45
Mountain, Albert VC 218
Mullins, Frank 161
Murray, Joseph 29-30
Murray, Tom 29
Murren 197
Murtagh, Fred 96
Murton 213

National Memorial Arboretum 41, 257-8
National Miners' Memorial 41, 257-8
National Unemployed Workers Movement 244
neurasthenia 200
Neuve-Chapelle 41, 55
Neuville St Vaast 67
New Delaval 22
New Rossington 157
New Stevenston 26
New Tredegar 148, 200, 202
Newby, Henry 68
Newhall Camp 33
Newport 139

Newcastle-upon-Tyne 22, 110, 115, 146, 176
Newsome, John 99
Nicholson, AD 26
Nicholson, Maurice 188
No Conscription Fellowship (NCF) 149
Noeux-les-Mines (cemetery) 64
Non-Combat Corps (NCC) 141, 149
Normansall, John 160
Normanton 60, 176
North Seaton 45
North Staffordshire Coal Owners 174
Northumberland Miners' Association 207
Norton-Griffiths, Sir John 59, 61-2, 66, 67, 120, 122, 154
Norwich 176
Not for Glory (book) 34
Nottingham Miners' Association 45, 50

Oakdale Colliery Company 198
O'Dell, Thomas 251
Ollivant, William 91
Onions, Robert 68
Orchard, Arthur 99-100
Ormskirk 35-6
Orwell, George 245
Oughtibridge Coal Company 214
Owen, Wilfred 15, 204-6

pacifists 98
Pals (battalions) 8, 18, 23, 39, 43, 48, 57, 107, 110, 153, 158, 236
Parfitt, Henry Thomas 166-7
Parkes, Walter 197-8
Parkinson, Sir Michael 6-7, 9
Parry, Daniel 209
Parry, Thomas 209
Parton, Alfred 238
Passchendaele 36, 96, 97, 104, 153-6, 177, 178, 179, 211, 213, 226-7, 228, 249
Peace Pledge Union 151
Pearce Register of British Conscientious Objectors (PRBCO) 146, 147, 149, 152
Pelham Committee (Work of National Importance) 140, 147
Pentonville (prison) 149
Pentwynmawr 198-9
Penygraig 199
Perry, Frank Depperriaz 65
Petit Boise 40, 123
Phillips, Robert 200-2
pigeons 28
'pit-brow-lasses' 11, 12, 13, 48
Plebs League 148
Poelcapelle (3rd Ypres) 96, 178
Poilcourt camp (prison) 192
Polygon Wood 29
Pontefract 23, 95, 150, 184, 219
Pontypridd 176, 208

Poperinge 67, 73, 121
Post Talbot 144
Portsmouth 81
Powell Duffryn and Bute Co 23, 200
Powis, John 150
Pozieres 160
Presbyterians 148
President II HMS 188
Prince, Len 258
Princetown 140, 144
Prisoner(s) of War (PoWs) 9, 102, 190-203, 240-1
Pritchard, Richard 99
Proto (rescue equipment) 67, 71, 123, 170-2, 174
Prowse, George VC 218
Puchevillers Military Cemetery 93

Quakers (see Society of Friends) 141, 145, 148-9
Quebec (Durham) 146
Quinn, John 88-9

Rackham, Harry 90
Radcliffe-on-Trent 99
Radstock 223, 224
Railway Wood 68-9, 167-9
Raley, Ald William Elmsley 32, 246
rationing 208
Rawmarsh 255
Red Cross 50, 102, 165, 235, 236, 237
Redmayne, Sir Richard 17, 43, 48, 49, 50, 189, 233
Reece, William Hayden 123
regiments/units:
    Argyle and Southerlands Highlanders 27
    Army Cyclists 188
    Army Service Corps (ASC) 52, 218
    Artists' Rifles 256
    Border 40, 42
    Coldstream Guards 39, 56, 60, 77, 79, 227
    Connaught Rangers 93
    Devonshire 199
    Duke of Cornwall's Light Infantry 136
    Durham Light Infantry (DLI) 8, 107, 108, 180, 218, 249
    East Anglians 90
    East Yorkshires 88, 90, 91-2, 95, 192
    Grenadier Guards 180
    Irish Guards 46, 180
    Highland Light Infantry 55, 56, 218
    Imperial Yeomanry 115
    Kings Liverpool 36
    Kings Own Rifle Corps 56, 178
    Kings Own Scottish 196
    Kings Own Yorkshire Light Infantry (KOYLI) 23, 33, 34, 38, 87, 89, 99, 100, 108, 110, 159, 180, 184-5, 193, 218
    King's Royal Rifles 215
    Lancashire Fusiliers 34, 54, 157
    Leicesters 192

Lincolns 194
London 199
Machine Gun Corps 35, 102
Manchesters 205
Monmouthshires 70, 107, 120
Norfolks 188
Northamptonshire 193
North Staffordshires 150
Northumberland Fusiliers (also see Tyneside Irish/ Scottish) 78, 89, 90, 107-8, 111, 113, 115, 118-21, 175-6, 218
Plymouth Battalion Royal Marines 79
(Prince Consort's Own) Rifle Brigade 216
Prince of Wales Leinster 218
Prince of Wales Own (West Yorks) 92-3, 180, 218
Royal Air Force 185-6
Royal Army Medical Corps (RAMC) 8, 16, 30-31, 52, 93-4, 107, 136, 146, 149, 170, 183, 195, 196, 222, 226, 236, 253
Royal Field Artillery 219, 221
Royal Flying Corps 181, 188
Royal Engineers (RE) 8, 15, 43, 57, 62-63, 65, 73, 90, 91, 99, 107, 117, 120-1, 123, 142, 159, 165, 167-9, 174, 198, 199, 209, 213, 221
(Tunnelling co's):
    170-177TCs 59
    170TC 61-2, 65, 69
    171TC 57-8, 69, 70, 120-2, 165, 209-10, 238
    172TC 69, 120, 122, 159, 198, 238
    173Tc 120
    175TC 67, 142
    177TC 67-9, 166, 169, 238, 240
    178-185TCs 59, 142
    179TC 71
    180TC 90
    182TC 239
    185TC 117
    250-58TCs 59
    250TC 120, 123
    252TC 200
    253TC 69, 199, 239
    254TC 69
    256TC 59, 69
    258TC 64
Royal Anglesey 73
Royal Field Artillery 22, 90, 100, 109, 215, 218
Royal Fusiliers 180
Royal Horse Artillery 251
Royal Marine Light Infantry 182
Royal Munsters 99, 101
Royal Naval Division 218
Royal Navy Air Service 188
Royal Scots 40, 55, 56, 86, 100, 194, 218, 228
Royal Sussex 63
Royal Welch Fusiliers 56, 61, 63, 137, 169, 180, 231
Scots Guards 180

Scottish Bantams 107
Sherwood Foresters 56, 102, 151, 159, 169, 177-78, 180, 182, 216, 218, 229
Somerset Light Infantry 74, 223
South Staffordshires 65
South Wales Borderers 21, 77, 120, 136, 167, 217
Tank Corps 107, 180
Territorials 37, 40, 42, 55, 70, 96, 136, 183
Tyneside Irish 111, 113, 118, 119
Tyneside Scottish (Northumberland Fusiliers) 1, 107-8, 11, 113, 115, 118, 120, 158, 175-6Welsh 20-21, 44, 107, 137
Welsh Fusiliers 167
Welsh Guards 176
West Riding 38
West Yorkshires 89, 99, 100, 248
Worcestershires 211, 256
York and Lancasters 19, 23, 27-8, 32-3, 34, 38, 42, 43, 57, 87, 89, 90-1, 94, 95, 96, 99, 101, 102, 104, 138, 150, 159-60, 181, 184, 186, 197, 220, 224, 225, 246, 248, 249
Requiem for Will (book) 73
Rhodes, John Harold VC 180
Rhondda 20, 23, 44, 199, 208
Rhondda, Lord – see Thomas, David A
Rhur 202
Richards, Frank 28-9, 231-2
Ripley, George 87
Risca 209
Roberts, Arthur Brinley 148
Roberts, Isaacs 226-7
Roberts, Richard 169
Roberts, Samuel MP 250
Robeson, Paul 80
Robinson, JK 26
Robson, Henry Howie VC 40
Roclincourt 175
Roclincourt Military Cemetery 162
Roehampton 235
Rolfe, Canon Thomas 157
Rollett, Ernest 100-01
Rotherham 34, 42, 83, 85, 93, 94, 96, 97, 100, 104, 109, 138, 150, 183, 187, 222, 252
Rotherham Advertiser 88, 89, 92, 93, 94, 95, 96, 100, 138, 187, 192, 234
Rouge Bancs 40
Royal Flying Corps 106
Royal Marines 81
Royal Navy 80, 87, 90, 136, 185
Royal Navy Volunteer Reserve 29
Royal School of Mines 188
Royston (Barnsley) 179, 180, 246, 247
Runciman, Walter 140
Rushton, Joseph 227-8
Russelbury, Harry 64
Russian Order of St George 102

Rutter, Rev William 157
Ryon, Pte Patrick 34-5
Ryton-on-Tyne 1, 110, 112, 113-14, 117

Sailly-sur-la-Lys Canadian Cemetery (Picardy) 90
Safety in Mines Research Board 122
Salvation Army 244, 247
Sambre-Oise Canal (Ors) 205
Sankey, John (Commission) 233, 243
Sanna-I-Yat 136
Sassoon, Siegfried 29, 153, 205
Sayers, William 99
Scarpe (battle) 160, 181
Schleswig camp (prison) 194
Scholes, Alfred 102
Scottish miners 46, 48, 56
Seaton Delaval 22
self-contained breathing apparatus (SCBA) 17, 67, 70, 123, 170-5
Senghenydd 73-6
Serre 27, 40, 107, 108, 109, 229
Sheffield 86, 93, 99, 104, 109, 160, 188, 237, 256
Shepherd, Albert Edward VC 178-80, 247
Shiel, Lieut FRG 22
Shilden 188
'Shot at Dawn' (memorial) 228-30, 257-8
Shrewsbury 150
Shuttington 65
Siebe Gorman 173
Silver War Badge 19, 31, 142, 223, 251
Simmes, Robert William 228-9
Sirhowy Valley 120
Skelton 93
Skipton 80
Smallman, William 142-3
Smillie, Alexander 148
Smillie, Robert 148, 189, 208, 233
Smith, George Deville 162
Smith, Rev James A VC 40
Smith, Harry 194
Smith, Herbert 251
Smith, William 188
Society of Friends (Quakers) 140, 145, 147, 148, 149
Somme 28, 33, 36, 39, 47, 71, 75, 78, 91, 93, 94, 96, 99, 100, 101, 106-43, 150, 153, 158, 160, 178, 179, 197, 200, 209, 212, 213, 214, 225, 227, 229, 258
Southmoor 67
South Shields 40
South Wales Argos (newspaper) 198
South Wales Miners' Federation (SWMF) 20, 140, 207, 242
South Wingate 47
South Yorkshire Times 195
Spanbroekmolen (Lone Tree Crater) 121, 123
Spence, Len 36-7
Spencer, James Hebert 70
Spencer, Leonard 38

Spennymoor 148
St George's Medal (Russian) 63
St John's Ambulance Association 16, 30-31, 50, 52, 115, 178, 183, 223, 226
St Peter's Church (Whinney Hill) 84, 87, 91-4, 96, 104
Stacey, Thomas 90
Star and Garter Home (Surrey) 221, 235
Star Medal (14-15) 37, 38, 66, 90, 182, 229, 239
Staveley Coal and Iron Company 177, 239
Steeland 48
Stella Coal Company 110, 115, 117
Stoke-upon-Trent 142
Stoke War Hospital 63
Stones, Robert 193
Straker, William 45, 207
Strangeways (prison) 149
Stone, Charles 218
Stones, John 183-4
Stourbridge 176
stretcher bearer(s) 94, 95, 96, 115, 150, 160, 195, 226-7
Streets, John William 27-8
Subterranean Sappers (book) 67
Sunderland 145
Sunken Road Military Cemetery 161
Sutherland, Sir William 250
Sutton-in-Ashfield 182-3
Swansea 139
Swinton (Rotherham) 188, 251, 256
Sykes, Charles 89

Tamworth 65
Tanfield 119
Tankersley 138, 174
tanks 75-6, 155-6
Tavistock Hospital 151
Taylor, Harry 94-6
Taylor WS 148
The National Archives (TNA) 190
The Proud Valley (film) 80
The Valley (book) 236
Thiepval Memorial 73, 92, 115, 117, 119
Thomas, Arthur 68
Thomas, David A (Lord Rhondda) 44, 156, 208
Thomas, JS 148
Thornhill, David 96
Thrybergh 83, 88, 90-4, 100, 183-4
Thurnscoe 36-7
Tidworth 183
Tilloy British Cemetery (Arras) 181
Tinsley 93
Tipton 137
Tomaselli, Phil 46, 55
Tow Law 68
Trafford, Richard 35-6
Tremble, Arthur 113, 115
tribute medals 31, 252-3

Index • **287**

Trones Wood 75
tunneller(s) 41, 57-62, 65, 67, 68-74, 107, 116, 122-3, 142, 153-4, 169, 198, 209-10, 238-40, 258
Turner, Arthur 182-3
Tyas, Clifford 32-3
Tyler, Charles 86
Tyneside Irish (book) 111
Tyneside Scottish (book) 111
Tyne Cot 154, 159

Upton, James VC 56

Vermellen 77
Vermelles 201
Victoria Cross (VC) 8, 39, 47, 53, 54, 55-6, 79, 95, 135, 136, 137, 175-80, 213-18, 247
Victory Medal 31, 38, 77, 90, 142, 182, 223, 229
Villers-Pouich 179
Vimy (Ridge) 67, 153, 175, 181, 239, 239

Wainwright, Gary 218
Wainwright, Harry 218-20
Wainwright, John William 102-3
Wakefield 24
Walker family (Barnsley) 142
Walker, Serg W 99
Walton (prison) 149
Walton, James MP 145
Wandsworth (prison) 149
Ward-Aldham, Julia 237
war memorial(s) 37, 46, 85-6, 112, 253-8
war hospitals 236
War Underground (book) 67
Warren, Francis Percy 42
Warrington 225
Watkins, Arthur 78-9
Watts, Enoch 92
Warne, George 148
Wath-upon-Dearne 159, 251
Welsh Miners' Federation (The Fed) 48
Wentworth 251
West Rainton 46
West Stanley 148
West Yorkshire Coal Owners' Association 23, 38
Western Front 34, 36, 40, 45, 59, 63, 65, 67, 71, 84, 106, 120, 122, 152, 153, 161-2, 164, 170, 174, 178, 191, 193, 197, 200, 202, 208, 210, 220, 228, 240, 256
Westlake, Donald Alexander 99
Whalley, Thomas 149
Wheatley Hill 47
Whitla, Sir William 240
Wigan 12, 41, 54, 80
Willert, George 92-3
Williams, Edward 148
Williams, John Henry VC 217-8
Willington 121

Wilson, George F VC 39
Wilson, JRR (mines inspector) 25-6, 139, 152
Wilton, Jessie 230
Wimereux Communal Military Cemetery 210
Winborn, WT 173
Windy Corner (Guards' Cemetery) 91
Wing, Arthur 249
Winter, John George (CO) 146
Wharncliffe War Hospital (Sheffield) 237, 238
Whinney Hill 84-5, 90, 96, 99, 100, 104
Whitby 197
Whitefield, George 149
Whitehaven 17, 26, 40, 43
Whitehaven Colliery Company 42
Whitwell 27
Wickersley 109
Wilkinson, Oliver 191
Winchester (prison) 149
Wolverhampton 242
Wombwell 78, 161, 220, 250, 254
Women's Peace Crusade 149
Wood, Ritchie 69-70, 166, 198
Woodbridge, Lesley 71
Woodcock, Thomas VC 180
Woodlands Model Village 111
Woolley, George Arthur 169
Woolley, Walter 92
Worksop 188
Wormwoods Scrubs (prison) 145, 148, 149, 150, 151
Worsbrough 162
Wrexham 71
Wright, George 100
Wright, George (CO) 146
Wright, James 196-7
Wright, J E Seaman-Gunner 8

Y-Farm Military Cemetery 42
Ynyshir 209
Yorkshire Miners' Association (YMA) 49, 64, 98, 145, 160, 164, 248, 249, 251
Youll, John VC 218
Young Thomas VC 47, 218
Ypres 29, 43, 46, 58, 60, 62, 67, 68, 73, 87, 104, 121, 122, 137, 141, 150, 154, 159, 167, 171, 177, 178, 179, 184, 198, 201, 210, 212, 215, 217, 228, 256
Yser Canal 68, 91
Ystradgynlais 226

Zillebeke 38, 68-9, 167
Zonnebke 211